C0-APY-316

Earthquake Resistant Construction of Gas and Liquid Fuel Pipeline Systems Serving, or Regulated by, the Federal Government

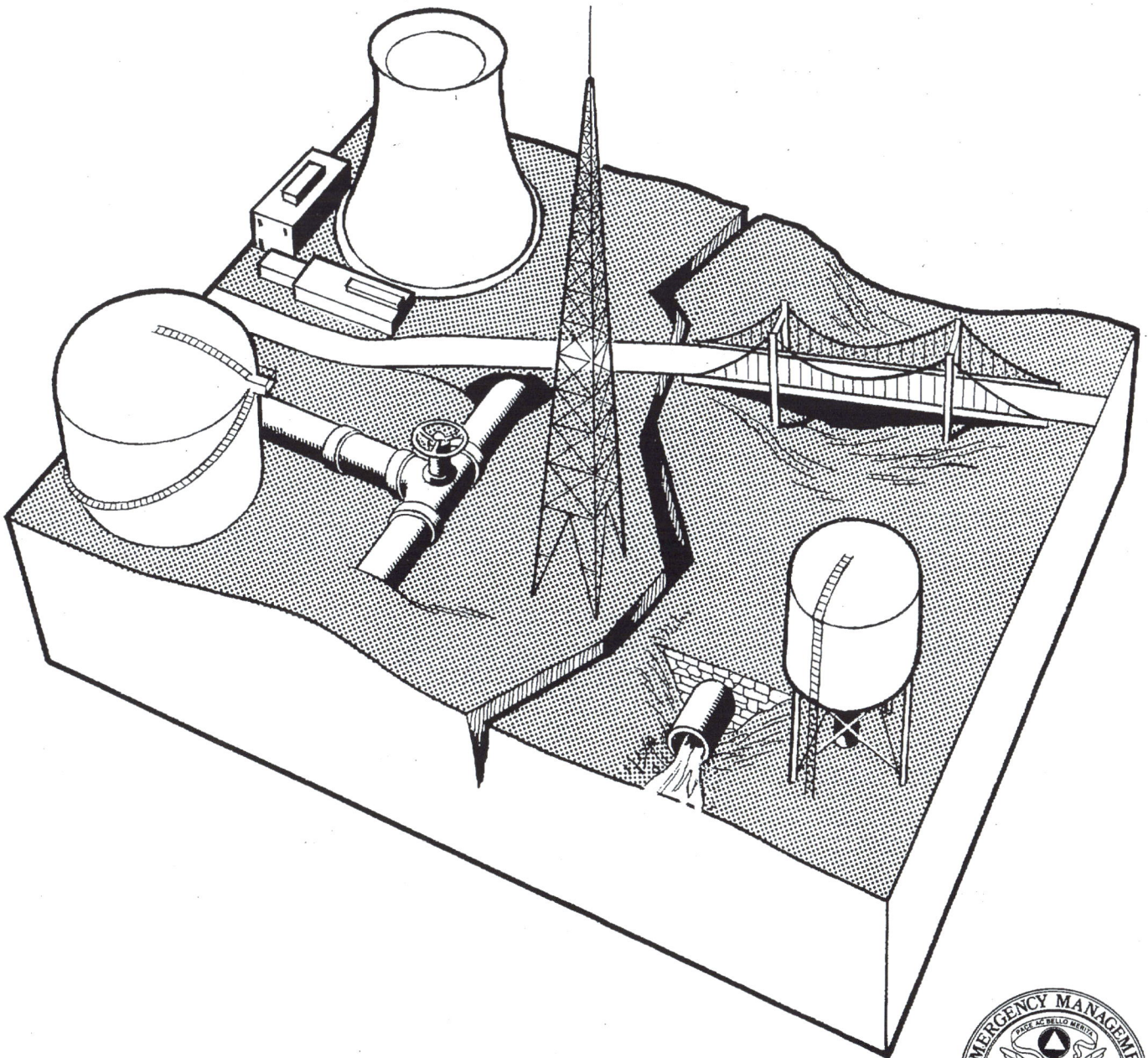

Issued in Furtherance of the Decade
for Natural Disaster Reduction

Earthquake Hazard Reduction Series 67

EARTHQUAKE RESISTANT CONSTRUCTION OF GAS AND LIQUID FUEL PIPELINE SYSTEMS SERVING, OR REGULATED BY, THE FEDERAL GOVERNMENT

Felix Y. Yokel
Robert G. Mathey

March, 1992
Building and Fire Research Laboratory
National Institute of Standards and Technology
Gaithersburg, MD 20899

U.S. DEPARTMENT OF COMMERCE
Barbara Hackmann Fraklin, *Secretary*
Technology Administration
Robert M. White, *Under Secretary for Technology*
National Institute of Standards and Technology
John W. Lyons, *Director*

Prepared for
Federal Emergency
Management Agency
500 C Street S.W.
Washington, DC 20472

ABSTRACT

The vulnerability of gas and liquid fuel pipeline systems to damage in past earthquakes, as well as available standards and technologies that can protect these facilities against earthquake damage are reviewed. An overview is presented of measures taken by various Federal Agencies to protect pipeline systems under their jurisdiction against earthquake hazards. It is concluded that the overall performance of pipeline systems in past earthquakes was relatively good, however, older pipelines and above-ground storage tanks were damaged in many earthquakes. Modern, welded steel pipelines performed well, however, damage occurred in areas of major ground displacements. Available standards and regulations for gas pipelines do not contain seismic provisions. Standards and regulations for liquid fuel pipelines contain only general references to seismic loads. Standards and regulations for above-ground fuel storage tanks and for liquefied natural gas facilities contain explicit seismic design provisions. It is recommended that a guideline for earthquake resistant design of gas and liquid fuel pipeline systems be prepared for Federal Agencies to ensure a uniform approach to the protection of these systems.

Key Words: codes; earthquake engineering; fuel pipelines; lifelines; liquefied natural gas; natural gas; oil; oil storage; pipelines; seismic design; fuel storage tanks.

TABLE OF CONTENTS

ABSTRACT iii

TABLE OF CONTENTS v

DEFINITIONS vii

LIST OF ACRONYMS viii

EXECUTIVE SUMMARY ix

1.   INTRODUCTION 1

2.   COMPONENTS OF NATURAL GAS AND LIQUID FUEL PIPELINE SYSTEMS 2
     2.1 General Description 2
     2.2 Gas Pipeline Systems 8
          2.2.1   Pipelines 8
          2.2.2   Compressor Stations 9
     2.3 Liquid Fuel Transmission Systems 9
          2.3.1   Pipelines 9
          2.3.2   Pumping Stations 10
          2.3.3   Storage Tanks 10

3.   PERFORMANCE OF GAS AND LIQUID FUEL PIPELINE SYSTEMS IN PAST
     EARTHQUAKES 11
     3.1 Pipelines 11
          3.1.1   Earthquake Effects 11
                  3.1.1.1   Traveling Ground Waves 11
                  3.1.1.2   Permanent Ground Displacements 11
                  3.1.1.3   Secondary Effects 12
          3.1.2   Factors Affecting Performance of Pipelines 13
          3.1.3   Failure Mechanisms 14
          3.1.4   Remedial Measures 16
          3.1.5   Summary 16
     3.2 Tanks 17
          3.2.1   Overall Performance Record 17
          3.2.2   Earthquake Effects Causing Failures of Tanks 17
          3.2.3   Factors Affecting Tank Performance 17
          3.2.4   Failure Mechanisms 18
          3.2.5   Design Methodologies 19
          3.2.6   Lessons Learned 19
     3.3 Structures and Above Ground Support Facilities 20
          3.3.1   Overall Performance Record 20
          3.3.2   Design Methodologies 21
          3.3.3   Lessons Learned 21

4. AVAILABLE DESIGN CRITERIA, REMEDIAL MEASURES, STANDARDS, AND DESIGN
GUIDES 22
   4.1 Introduction 22
   4.2 Design Criteria 22
      4.2.1 Development of Design Criteria 22
      4.2.2 Current Design Criteria 23
         4.2.2.1 Pipelines 23
         4.2.2.2 Storage Tanks 24
         4.2.2.3 Structures and Support Facilities 26
   4.3 Emergency Response, Evaluation, Repair and Retrofitting 26
      4.3.1 Emergency Response 26
      4.3.2 Evaluation 28
      4.3.3 Repair 28
      4.3.4 Retrofitting 28
   4.4 Standards, Codes, and Design Guides 30
      4.4.1 Summary of Available Codes and Standards 30
         4.4.1.1 Codes and Standards for Pipelines 30
         4.4.1.2 Seismic Design Provisions in the Codes and Standards
              for Pipelines 30
         4.4.1.3 Codes and Standards for Storage Tanks 32
         4.4.1.4 Seismic Design Provisions in the Codes and Standards
              for Storage Tanks 32
         4.4.1.5 Codes and Standards for Structures and Support
              Facilities 33
         4.4.1.6 Seismic Design Provisions in the Codes and Standards
              for Structures and Support Facilities 33
      4.4.2 Comments on Available Codes and Standards 34
         4.4.2.1 Siting 34
         4.4.2.2 Pipelines 35
         4.4.2.3 Storage Tanks 36
         4.4.2.4 Structures and Facilities 36
   4.5 Summary 37

5. FEDERALLY CONTROLLED SYSTEMS 37
   5.1 Introduction 37
   5.2 Federal Practices 38

6. SUMMARY AND RECOMMENDATIONS 45
   6.1 Summary of Findings 45
      6.1.1 System Vulnerability 45
      6.1.2 Remedial Measures 45
      6.1.3 Existing Guidelines and Standards 46
      6.1.4 Federal Practices 47
   6.2 Recommendations 47

7. ACKNOWLEDGMENT 48

8. REFERENCES 49

APPENDIX   DISCUSSIONS WITH STAFF MEMBERS FROM FEDERAL AGENCIES 61

DEFINITIONS

To assist the reader with the interpretation of terminology used in this report, a list of definitions is given below. Most of these definitions were taken from the Code of Federal Regulations, 49 CFR, Parts 192, 193, and 195 and from the technical literature. Some of the components of pipeline systems are also described in the text of the report.

- **Distribution line:** a pipeline other than a gathering or transmission line.

- **Gas:** natural gas, flammable gas, or gas which is toxic or corrosive.

- **Gathering line:** a pipeline that transports gas from a current production facility to a transmission line or main, or a pipeline 203 mm (8 in) in nominal diameter that transports petroleum from a production facility.

- **Liquefied natural gas (LNG):** natural gas or synthetic gas having methane (CH_4) as its major constituent which has been changed to a liquid or semisolid.

- **Liquid fuel:** crude oil and petroleum products.

- **Main:** a distribution line that serves as a common source of supply for more than one service line.

- **Pipe:** any pipe or tubing, usually cylindrical, used in the transportation of gas or liquid fuel, including pipe-type holders.

- **Pipeline:** any pipe or tubing, and associated joints, welds, couplings, tees, bends, and appurtenances, through which gas or liquid fuel move in transportation, excluding facilities to which the pipeline is connected, such as compressor units, metering stations, pumping stations, etc.

- **Pipeline facility:** new and existing pipelines, rights-of-way, and any equipment, facility, or building used in the transportation of gas or liquid fuels, or in the treatment of gas during the course of transportation.

- **Pipeline systems** (as defined in this report): all facilities and components that are needed for the transportation, distribution, and storage of natural gas, crude oil, and petroleum products.

- **Service line:** a distribution line that transports gas from a common source of supply to (a) a customer meter or the connection to a customer's piping, whichever is farther down stream, or (b) the connection to a customer's piping if there is no customer meter that measures the transfer of gas from an operator to a customer.

- **Storage tank:** a container for storing gas or liquid fuels, including an underground cavern.

- **Transmission line:** a pipeline through which gas and liquid fuels are transported from source areas to distribution points, processing plants, or storage areas.

- **Transportation of gas:** the gathering, transmission, or distribution of gas by pipeline, or the storage of gas, in or affecting interstate or foreign commerce.

LIST OF ACRONYMS

ASCE	American Society of Civil Engineers
ASME	American Society of Mechanical Engineers
ANSI	American National Standards Institute
API	American Petroleum Institute
AWWA	American Water Works Association
BSSC	Building Seismic Safety Council
CFR	Code of Federal Regulations
CPUC	California Public Utility Commission
DOA	Department of Agriculture
DOD	Department of Defense
DOE	Department of Energy
DOI	Department of Interior
DOT	Department of Transportation
EERI	Earthquake Engineering Research Institute
EPA	Environmental Protection Agency
FEMA	Federal Emergency Management Agency
FERC	Federal Energy Regulatory Commission
FHWA	Federal Highway Administration
GLFL	Gas and Liquid Fuel Lifeline
GSA	General Services Administration
HUD	Department of Housing and Urban Development
LNG	Liquefied Natural Gas
LPG	Liquid Propane Gas
MMS	Minerals Management Service of DOI
NASA	National Aeronautics and Space Administration
NCEER	National Center for Earthquake Engineering Research
NIST	National Institute of Standards and Technology
NOAA	National Oceanic and Atmospheric Administration
NTSB	National Transportation Safety Board
OCS	Outer Continental Shelf
PGE	Pacific Gas and Electric Company
TCLEE	Technical Council on Lifeline Earthquake Engineering, ASCE
TRB	Transportation Research Board
TVA	Tennessee Valley Authority
UBC	Uniform Building Code
USGS	U.S. Geological Service of DOI

EXECUTIVE SUMMARY

The vulnerability of gas and liquid fuel pipeline systems to damage in past earthquakes, as well as available standards and technologies that can protect these facilities against earthquake damage are reviewed. Gas and liquid fuel pipeline systems considered include all facilities and components that are needed for the transportation, distribution, and storage of natural gas, crude oil, and petroleum products. An overview is presented of measures taken by various Federal Agencies to protect fuel pipeline systems under their jurisdiction against earthquake hazards.

It is concluded that the overall performance record of gas and liquid fuel pipeline systems in past earthquakes was relatively good. However, catastrophic failures did occur in many earthquakes, particularly in areas of unstable soils.

Modern, welded ductile steel pipelines, with adequate corrosion protection, have a good performance record. Failures that did occur were mostly caused by large, permanent soil displacements. Older pipelines, including welded pipelines built before 1950 in accordance with quality control standards less stringent than those used currently, as well as segmented cast iron pipelines, have been severely damaged. Some pipeline failures were attributable to the collapse of supporting structures to which they were attached.

Above-ground storage tanks, particularly those with large height-to-width ratios, were damaged in many earthquakes. Damage was caused by buckling and rupture of the shell, inadequate anchorage, excessive foundation settlements, inadequate flexibility of pipe connections, and contact of the sloshing liquid with the roof structure.

Pumping and compressor stations generally have performed well. Other above-ground support facilities, which were designed to resist earthquakes, suffered only limited structural damage. In many instances, however, problems were caused by inadequate tiedown of equipment and anchorage to the supporting foundations. Equipment outages were often caused by falling debris, collision with other items, sliding furniture and other objects, or failure of electrical supplies.

Fuel pipeline systems can be designed to be protected against most earthquake hazards. One of the most efficient and economical ways to obtain earthquake protection for new facilities is proper siting. Storage tanks and other above-ground facilities can normally be located to minimize exposure to unstable ground. Transmission and distribution pipelines traverse large areas and must often cross zones of potentially unstable soils. Nevertheless, careful planning in route selection, pipeline orientation, and location of critical components can promote good performance during earthquakes.

In addition to proper siting, pipeline systems can be designed to resist most, but not all potential earthquake loads and displacements. Criteria and guidelines for pipeline system design were presented by the American Society of Civil Engineers (ASCE, 1984). Criteria for tanks and other structures are incorporated in many existing standards. An effective protection against the environmental consequences of storage tank failures can be provided by secondary containment using earth dikes. Such secondary containments are presently required only for liquefied natural gas (LNG) storage facilities.

For existing facilities, retrofit and replacement of older facilities in critical areas should be considered. Methods for inspecting and retrofitting older pipelines are available.

Present standards for pipelines generally do not adequately address the earthquake problem. Neither the pipeline standards, nor the standards for oil storage tanks address the need for siting studies, even though such studies are often performed in practice. This deficiency could have adverse consequences, particularly in the Central and Eastern U.S., where the need for earthquake resistant design is not always fully recognized. There are also no secondary containment requirements for liquid fuel storage tanks, even if these tanks are located in environmentally sensitive areas. Standards and Federal regulations for LNG storage facilities contain siting criteria, secondary storage provisions and lateral force design requirements.

Three Federal Agencies have regulatory responsibilities for pipeline fuel transportation systems: The Department of Transportation (DOT) regulates oil and gas pipelines; the Federal Energy Regulatory Commission (FERC) regulates oil and gas pipelines and all LNG facilities, including terminal and storage facilities; and the Department of Interior's Minerals Management Service (MMS) regulates offshore production and transmission facilities. To some extent, the responsibilities of these agencies overlap. The review and approval of facilities by these agencies are based on the relevant provisions in the Code of Federal Regulations (CFR) and on engineering judgment. Explicit requirements for geological and seismological studies, secondary storage, and earthquake resistant design are included in the federal regulations for LNG facilities. The federal regulations for gas pipelines, as well as other standards referenced in these regulations, do not address earthquake resistant design. The federal regulations for liquid fuel pipelines have only a very general requirement for earthquake resistant design which is not sufficiently detailed or focused to provide direction on the critical aspects of seismic performance. Commercial standards, which address the earthquake resistant design for liquid fuel storage tanks, are adopted by reference.

Most other federal agencies do not own and operate pipeline systems (except for relatively short pipelines), but many agencies own distribution systems and storage facilities. Most agencies address the earthquake problem in some way, but there is no uniform approach to the protection of gas and liquid fuel pipeline systems against earthquake damage among agencies, and sometimes within agencies.

It is recommended that a guideline for earthquake resistant design of oil and liquid fuel pipeline systems be prepared for Federal Agencies to ensure a uniform approach to earthquake resistant practices by all Agencies. This guideline should adopt existing standards and regulations by reference, but add requirements for siting, and for secondary storage for some above-ground liquid fuel tanks.

Since the proposed federal guideline may eventually result in an updating of present federal regulations, close coordination between FEMA, DOT, FERC, and MMS, as well input from industry, will be required.

1. INTRODUCTION

As part of the Action Plan for the Abatement of Seismic Hazards to Lifelines (Building Seismic Safety Council, 1987), the National Institute of Standards and Technology (NIST) reviewed measures presently taken by Federal Agencies to protect gas and liquid fuel pipeline facilities against seismic hazards. This report summarizes the results of the study. The study deals with pipeline systems for oil, other petroleum products, and natural gas. Gas and liquid fuel pipeline systems consist of all facilities and components that are needed for the transforation, distribution, and storage of natural gas, crude oil, and petroleum products. The study does not deal with oil and natural gas production facilities, oil refining facilities, rail transmission, and pipeline transmission of coal slurries. A similar study has been conducted by NIST for electrical transmission and telecommunication facilities (Yokel, 1990)

All privately owned fuel transmission pipeline systems constructed in the U.S. must comply with Title 49, Transportation, Parts 190, 191, 192, 193, and 195 of the Code of Federal Regulations, (49 CFR, 1990) which deal with the transportation of natural and other gases, liquefied natural gas facilities, and the transportation of hazardous liquids by pipelines. This code provides minimum safety standards for gas and fuel transmission and storage facilities in the United States. While private industry must comply with the code, the provisions are not mandatory for Federally owned or operated lines.

Available data on earthquake damage of oil and gas pipeline systems and related facilities from the United States and many other countries indicate that earthquakes pose one of the major threats to pipeline operations, and their effects must therefore be adequately accounted for in the design process (ASCE, 1983).

Approximately one half of the nation's supplies of crude oil and petroleum products, and virtually all of its natural gas supplies, are transported through a network of 2.7 million kilometers (1.7 million miles) of pipelines (TRB, 1988). These pipelines provide a vital transportation service and extend over long distances and traverse a variety of different soil and geologic conditions, as well as regions with different seismicities. Thus, they are exposed to a wide range of ground conditions and behavior. Pipelines are interconnected with other pipelines, storage structures, and support facilities. Damage in one part of these complex systems can have important repercussions on the flow and serviceability in the other parts of the systems.

In the seismic design of buildings, bridges, and other structures above ground, inertial force is usually the most important factor to consider (Singhal & Benavides, 1983). Burial of pipelines tends in general to isolate them from the effects of inertial forces, but makes them susceptible to relative ground motions which cause distortions and strains (Hall, 1987). Large, permanent ground movements in the form of surface faulting, soil liquefaction, and landslides, are the most troublesome sources of damage to gas and liquid fuel pipelines (O'Rourke, 1987). A critical aspect of earthquake engineering for pipeline systems is understanding the properties of surrounding soils and the potential reactions of these soil deposits to earthquake excitation. This requires input from seismologists, geologists and geotechnical engineers which is not explicitly required in the Federal Regulations and not always provided in present practice.

1

Seismic damage to underground piping systems has been caused by fault displacements, landslides, liquefaction of sandy soils and associated lateral spreading and earthquake-induced settlements, differences in dynamic properties of two horizontally adjacent soil layers, and ground strains associated with traveling seismic waves. (Lee and Ariman, 1985, Ariman, 1987, O'Rourke, 1988, O'Rourke and Ayala, 1990).

Section 2 of this report contains a description of fuel pipeline systems and their components; the performance of fuel systems during past earthquakes is discussed in Section 3; Section 4 lists available standards, design guides, and earthquake damage mitigation technologies; Section 5 provides information on Federal practices; Section 6 contains conclusions and recommendations. Appendix A contains a summary of statements made by persons from various Federal Agencies and other organizations contacted in this study. A list of definitions is provided on page vii to assist the reader in the interpretation of terminology used in the report.

2. COMPONENTS OF NATURAL GAS AND LIQUID FUEL PIPELINE SYSTEMS

2.1 General Description

Natural gas and liquid fuels are conveyed through transmission pipelines from source areas to distribution points or processing plants. Gas and liquid fuel pipeline systems consist of all facilities and components that are needed for the transforation, distribution, and storage of natural gas, crude oil, and petroleum products. The major components of a typical natural gas production and transmission system are shown in the schematic drawing in figure 2.1 and in the plan of a specific system shown in figure 2.2. The components include pipelines, compressor stations, gas storage facilities, including underground storage fields, liquefied natural gas (LNG) storage facilities, and other storage facilities, LNG terminals, production and processing facilities, metering and control facilities, and distribution systems. Figure 2.3 is a schematic drawing of a petroleum transmission system, and figure 2.4 shows the layout of a particular U.S. pipeline system. Major components include, tank farms, oil field facilities, pumping stations, pipelines for crude oil and refined products, and monitoring systems. Gas and liquid fuel facilities include other components such as valves, regulators, communication and control systems, and maintenance facilities.

Control systems and communications are critical for safe and continuous conveyance of both gas and liquid fuels, and are vital for emergency response. They are also among the most vulnerable components of gas and liquid fuel facilities. Examples of critical components include monitoring instrumentation, communications equipment, computer hardware, remote valve controls, auxiliary equipment, emergency power systems, and uninterruptible power supplies (Nyman, 1991).

As an example of an automated monitoring and control system, an oil pipeline configuration in Florida, described by McPartland (1988), is shown in figure 2.5. This latter system is in an environmentally sensitive area and has a leak sensing system which will activate an alarm and effect an automatic shutdown when a leak is detected. It includes storage tanks, pumping stations, injection stations, booster stations, a port-dispatch station, check valves, generators, a control and monitoring system, and communication capabilities. Typical parameters that

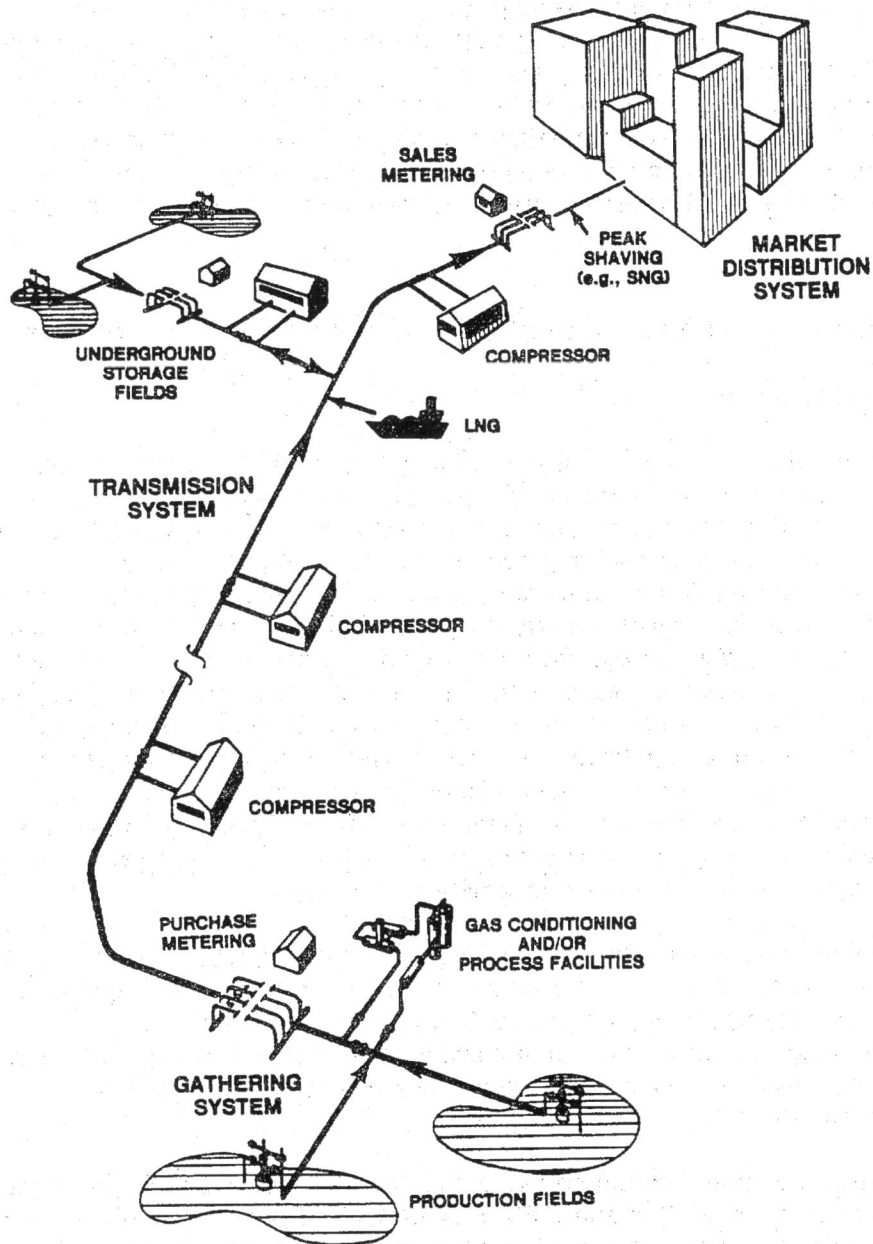

Figure 2.1   Schematic Drawing of Natural Gas Pipeline System (taken from National Petroleum Council, 1989)

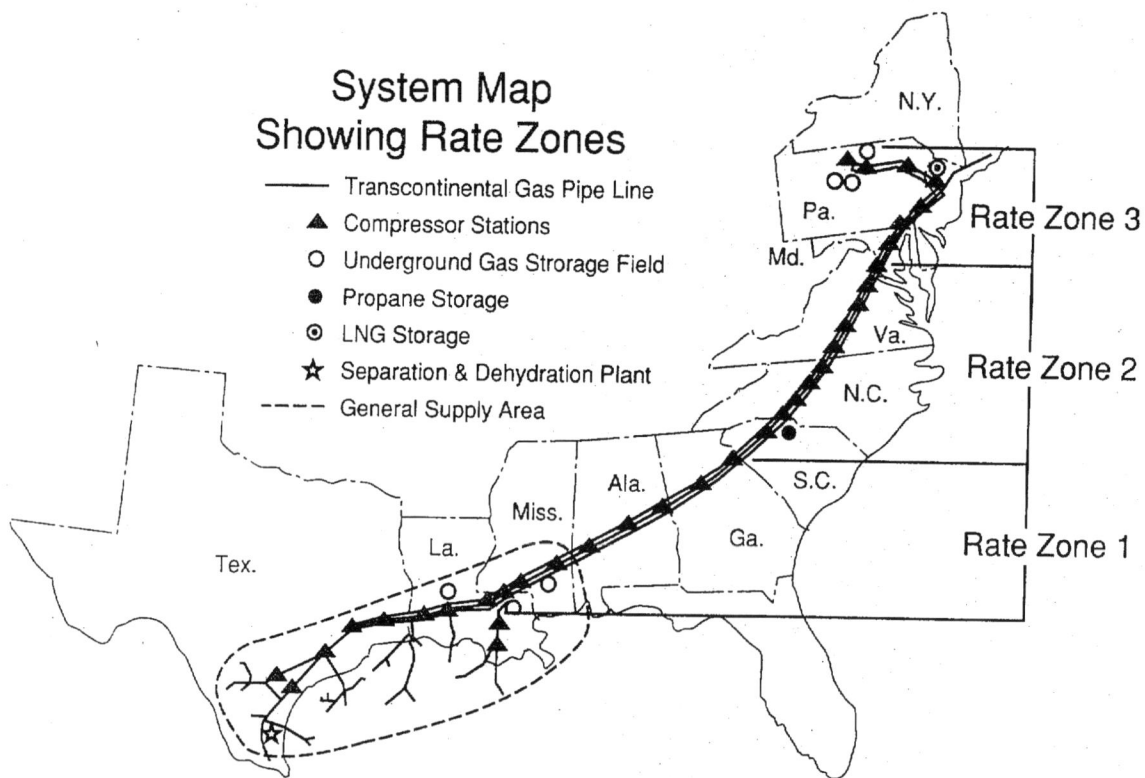

System Map
Showing Rate Zones

—— Transcontinental Gas Pipe Line
▲ Compressor Stations
○ Underground Gas Strorage Field
● Propane Storage
◉ LNG Storage
☆ Separation & Dehydration Plant
---- General Supply Area

Rate Zone 3

Rate Zone 2

Rate Zone 1

N.Y.

Pa.

Md.

Va.

N.C.

S.C.

Ga.

Ala.

Miss.

La.

Tex.

Figure 2.2   Map of Gas Pipeline System in the Southeastern United States (provided by the American Gas Association).

4

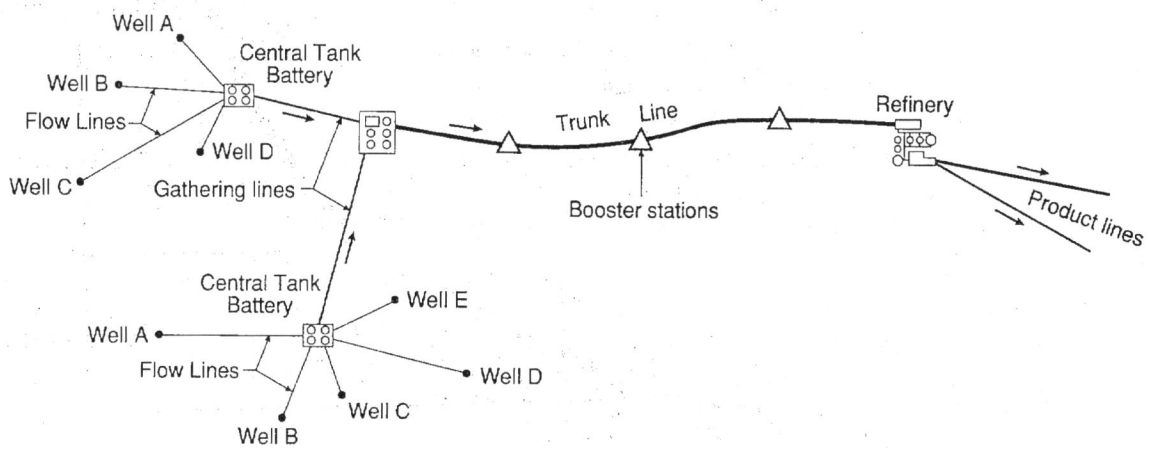

Figure 2.3 Schematic Drawing of Oil Pipeline System (taken in part from Giuliano, 1981).

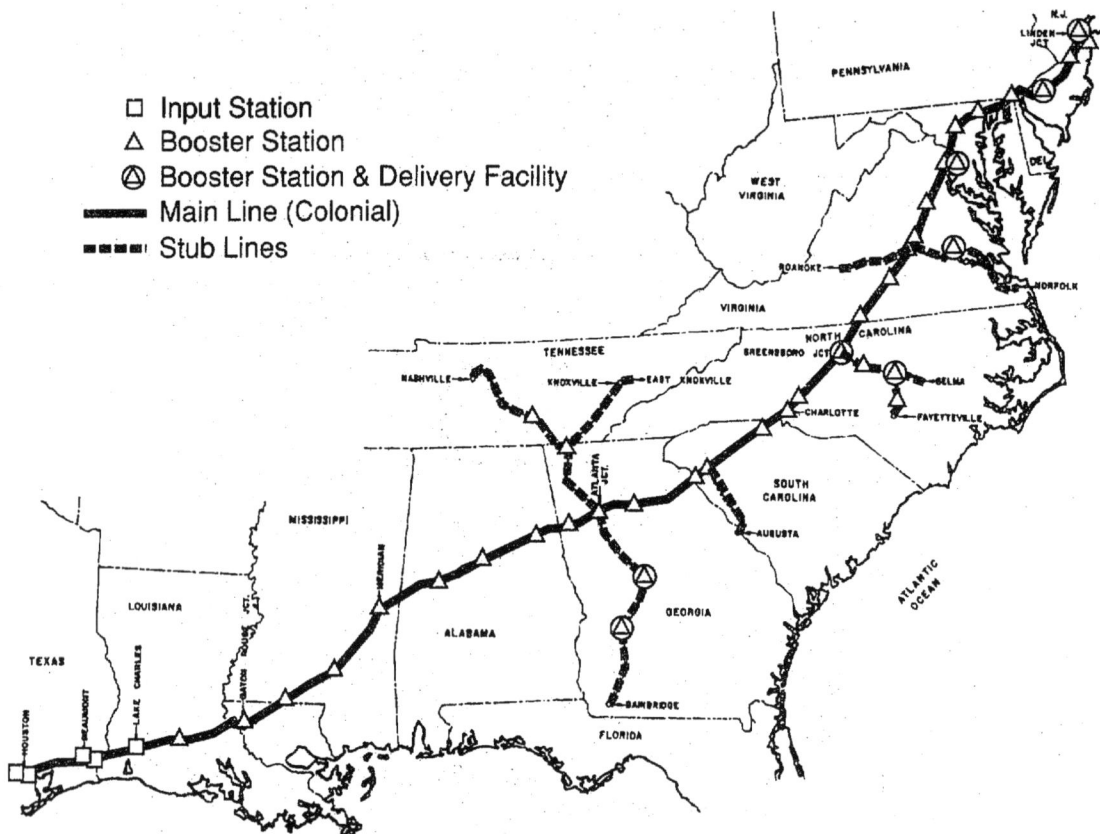

Figure 2.4   Map of Liquid Fuel Pipeline System in the Southeastern United States (taken from Giuliano, 1981).

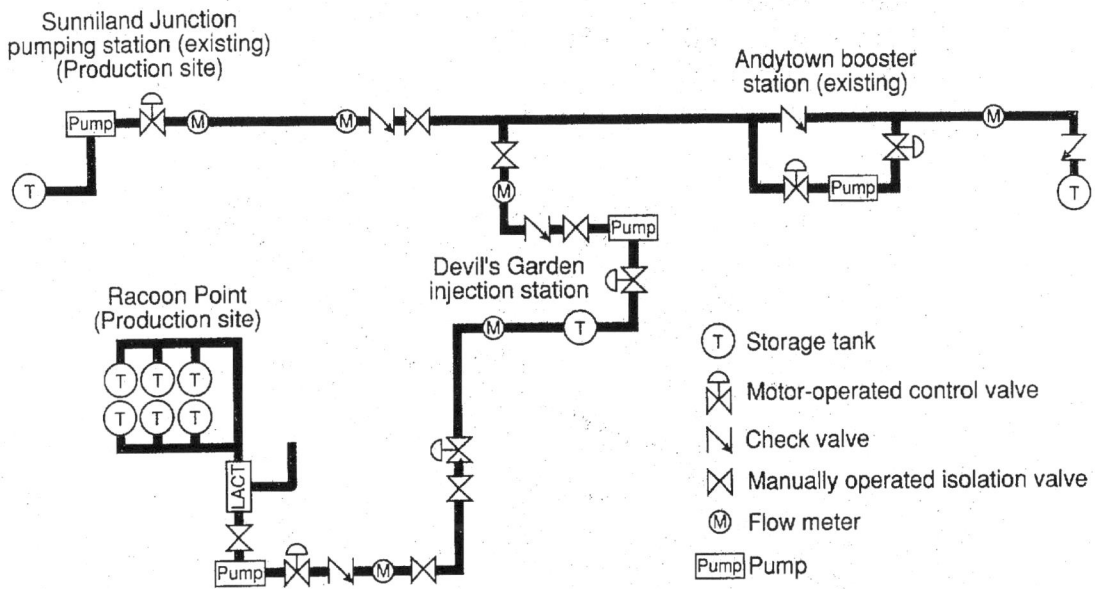

Figure 2.5   Pipeline System with Leak Detection System and Automated Shutdown Controls (taken from McPartland, 1988).

are measured to obtain information on the condition of a pipeline are pressure, flow rate, and linevolume balance. In certain situations, such as loss of pressure or failure of an operator to acknowledge a serious alarm within a preset time period, pipeline system shutdown is initiated automatically (Nyman, 1987). Natural gas systems require compressor stations and pressure regulators, and liquid fuel systems require pumping stations and sometimes heating and pressure reducing stations at intermediate locations.

According to Eguchi (1987), both gas and oil transmission systems should include isolation valves so that the effects of potential pipeline breaks can be confined to a relatively small portion of the system.

Because most pipelines have been constructed without special attention to seismic design, it is important that pipeline operations and maintenance programs begin to deal with potential earthquake damage and service interruptions (Nyman, 1987). McNorgan, 1973, reported that modern pipeline technology can provide earthquake resistant, but not earthquake proof structures.

2.2 Gas Pipeline Systems

2.2.1 Pipelines

Modern gas transmission pipelines are usually made of continuous girth-welded steel pipe and have diameters which range from 50 to 1060 mm (2 to 42 in), with most larger than 300 mm (10 in). These pipelines often carry internal pressures of 1.4 to 8.3 MPa (200 to 1,200 psi). Distribution pipelines have smaller diameters than transmission pipelines. The diameters of distribution pipelines are generally between 50 and 500 mm (2 and 20 in) and they may be composed of steel, cast iron, ductile iron, or plastic (EERI, 1986). Internal pressures in distribution pipelines generally range from 2 to 700 kPa (0.3 to 100 psi). Service pipelines have diameters generally less than 50 mm (2 in) and may consist of steel, copper, or plastic. Approximately 80 percent of all new distribution piping is made of plastic (EERI, 1986).

Hereafter are two examples of modern natural gas transmission facilities described in the literature. Fritsche, 1988, reported on a planned 238 km (148 mi) line in the state of Virginia. This line will include 43 km (27 mi) of 610 mm (24 in) pipeline near Leesburg, Virginia that will connect with the Virginia Natural Gas pipeline. This section of the pipeline will include a 10-MW (8,000-hp) compressor station and a measurement and regulating station. Design operating pressure of the line will be 6.9 MPa (1,000 psi) with actual delivery pressures of between 1.72 and 2.75 MPa (250 and 400 psi).

An example of a gas transmission system which includes liquefied natural gas (LNG) facilities and unloading terminals is given by Francis, 1984. He describes a natural gas transmission system operated by the British Gas Corporation, which includes compressor stations, LNG storage (above ground), LNG storage with a liquification plant (above ground), underground storage (salt cavity), and an exporting and reception terminal. There are about 120 installations in the U.K. national networks where gas is passed from national to regional control. These offtakes include compressor stations, storage facilities, and coastal terminals. Each of these offtake sites, and also some major junctions, are instrumented and the information telemetered to the data reduction computer at the regional center.

2.2.2 Compressor Stations

Compressor stations are generally located along a gas transmission line. Each station contains one or more centrifugal or reciprocating compressor units, and auxiliary equipment for purposes such as generating electricity, cooling discharge gas, and controlling the station (Tsal et al., 1986). It is possible to use two or more compressors at a station in parallel or in series.

Gas pipeline design discussed by Tsal et al (1986) involves: determination of the optimal number of compressor stations and their locations and design; and selection of the optimal pipe diameter and maximum allowable operating pressure. The following design variables need to be determined: number of compressor stations, compressor station locations, lengths of pipeline segments between compressor stations, diameters of pipeline segments, and suction and discharge pressures at each compressor station (Edgar et al., 1978).

2.3 Liquid Fuel Transmission Systems

2.3.1 Pipelines

Liquid fuel transmission pipelines often move large quantities of crude oil and petroleum products across active seismic regions. The Trans-Ecuadorian pipeline and the Trans-Alaskan pipeline are examples of major pipelines which transport oil across areas of high seismic activity (Anderson and Johnson, 1975). Anderson (1976), describes a crude oil pipeline system as consisting of three basic "segments". The primary segment is the line pipe that carries the crude oil. Piping associated with pump stations located along the length of the pipeline represents the second segment. The third segment is the piping associated with the tank terminal which is located at the end of the pipeline. All of these piping segments must be designed to resist forces developed during an earthquake. Pipelines may also be exposed to significant thermal forces and pressure forces which occur due to the flow of the oil and changes in direction. During operation, the pipeline is heated to a fairly constant temperature by the flow of oil in the pipe. Under operating conditions, the magnitude of the thermal forces is governed by the lowest temperature which occurred during installation of the pipeline.

Crude oil pipelines are usually buried below ground for economic, aesthetic, safety, and environmental reasons. In some instances, however, above ground support may be required, such as the Trans Alaskan pipeline, where structural support is needed to offset potentially large settlements caused by the melting of permafrost. Piping associated with a tank terminal is always located above ground (Anderson and Singh, 1976). Above ground support systems must allow movement due to thermal and pressure forces and also resist seismic forces. They offer the advantage of being readily accessible, either during normal operation or following a seismic disturbance. Above-ground pipeline systems are usually supported on gravel berms or pile bents.

An example of a buried pipeline, the Qinhuangdao-Beijing, China oil pipeline, which suffered some damage in the Tangshan earthquake, was described by Guan-Qing (1980). This oil pipeline went into operation in June 1975 and was 529 mm (20 in) in diameter, with a 7 mm (0.28 in) wall thickness, and was approximately 350 km (217 mi) long. The pipeline transported crude oil and was operated at a pressure of 5.9 MPa (856 psi) with the oil inlet temperature at 65 to 70° C (149 to 158° F). Along the pipeline length were two pump/heating stations and three heating stations. The pipeline was buried underground, except for river crossings, with 1.2 meter (47 in) cover. In most cases the pipeline was buried below freezing level and above the water table. The line used arc-spiral-welded steel pipe with a yield strength of 343 MPa (50 Ksi) and ultimate strength of 510 MPa (74 Ksi). The complete system was coated with reinforced asphalt and protected cathodically.

The Trans-Alaskan pipeline is 1230 mm (48 in) in diameter, with a wall thickness of 12 mm (0.462 in) or 14 mm (0.562 in). It is specially coated and cathodically protected from corrosion. It is 970 km (603 mi) long and crosses rivers and mountains, and reaches an elevation of 1460 m (4790 ft) (Factor and Grove, 1979). The line has eight pump stations. Slightly less than half of the pipeline is buried in stable soil. Above ground pipelines in areas of permafrost are insulated and jacketed with galvanized steel and mounted on crossbeams supported by vertical members set in the ground. The pipeline has 151 valves, including check valves to prevent a reversal of flow where oil is pumped uphill.

2.3.2 Pumping Stations

Friction loss associated with the flow of oil diminishes pressure in the pipeline. At certain intervals the pressure must be boosted by pump stations. Pumping is also required to transport oil uphill wherever this is required by topographic conditions. The spacing of these pump stations depends on the type of oil transported, the size of the pipe, and the topography. Pump station spacing generally ranges from 65 to 240 km.

2.3.3 Storage Tanks

Modern oil and liquid fuel storage tanks included in lifeline systems vary from 12 to 76 m (40 to 250 ft) in diameter with heights that are nearly always less than the diameter (Nyman, 1987). Ground supported tanks can be classified as anchored or unanchored tanks depending on their support conditions (Haroun, 1983). Most modern oil storage facilities use floating roof welded steel tanks (Kennedy, 1979).

Liquid fuel and gas storage tanks come in a variety of configurations. They may be elevated, ground supported, or partly buried. Ground supported, circular cylindrical tanks are more numerous than any other type because they are simple in design, efficient in resisting primary hydrostatic pressure, and can be easily constructed (Haroun, 1981).

3. PERFORMANCE OF GAS AND LIQUID FUEL PIPELINE SYSTEMS IN PAST EARTHQUAKES

Throughout the world, earthquakes have caused significant damage to underground pipelines, oil storage tanks, and some pump facilities (Eguchi, 1987). For above-ground components of pipeline systems, such as buildings and storage tanks, inertial forces resulting from ground shaking are a major concern. For buried pipelines, inertial forces are of little concern, but faulting, landslides, and liquefaction pose major problems (Hall, 1987).

3.1   Pipelines

3.1.1   Earthquake Effects

Ground deformations and displacements, rather than inertial forces caused by ground accelerations are the major cause of earthquake damage to pipelines. Ground deformations can be viewed as falling into two categories: ground strains caused by seismic wave propagation which do not result in large permanent deformations; and ground displacements caused by faults, soil liquefaction, settlements, and landslides. In addition to ground deformations, pipelines can also be damaged by secondary earthquake effects, such as failure of adjacent or connected structures, flooding, explosions and fires, and failure of support facilities.

3.1.1.1   Traveling Ground Waves

Hall reported in 1987 that there was no case of a modern buried welded steel pipeline failure attributable to ground shaking. However, one documented case of traveling ground wave damage to a corrosion-free modern continuous steel pipeline occurred to a 1067-mm (42-in) line during the 1985 Michoacan earthquake (O'Rourke and Ayala, 1990).

In other cases, pipeline damage from traveling ground waves has been observed in natural gas pipelines which were weakened either by corrosion or welds of poor quality (EERI, 1986). Recent Mexico City and Whittier, California earthquakes have shown that buried water pipelines were apparently damaged solely by seismic shaking effects since no large fault movement or soil liquefaction was found in either of the cities after the earthquakes (Wang, 1988). According to Wang, damage occurred mostly in regions of discontinuities in subsurface conditions. O'Rourke and Ayala, 1990, also reported cases in the 1964 Puget Sound, 1969 Santa Rosa, and 1983 Coalinga earthquakes where the sole damage mechanism appeared to be seismic wave propagation which did not result in permanent ground displacements.

3.1.1.2   Permanent Ground Displacements

Large permanent ground movements caused by surface faulting, soil liquefaction, and landslides are the most troublesome sources of earthquake damage to gas and liquid fuel pipelines (O'Rourke, 1987, EERI, 1986, Guan-Qing, 1980, Anderson, 1985, O'Rourke and Trautmann, 1981). Therefore, a primary concern for buried pipelines is their ability to accommodate abrupt ground distortions or differential displacements (ASCE, 1984). The amount and type of ground displacement across a fault or fault zone is one of the most important factors to be considered in the seismic design of pipelines crossing active faults (ASCE, 1983). Since ground displacements are in most cases difficult to predict, it is also difficult to develop designs which will protect pipelines against their effects. The most common forms of ground displacements are faulting, lateral spreading caused by liquefaction, and slope failures (landslides).

Pipelines that can otherwise sustain strong levels of shaking can be damaged severely by ground failures and local concentrations of movement (O'Rourke and Trautmann, 1981). Evidence reported in the literature indicates that underground pipelines perform worse in areas experiencing significant permanent displacement or ground failure (Eguchi, 1987, Lee and Ariman, 1985). It was pointed out that evidence from the 1906 San Francisco, 1952 Kern County, 1964 Niigata, 1964 Alaska, 1971 San Fernando, 1978 Miyagi-ken-oki, and 1983 Nihonkai-Chuba earthquakes shows an unmistakable correlation between permanent ground displacement and buried pipeline damage (O'Rourke, 1987, EERI, 1986). Ariman, 1984, noted from detailed examination of records from the 1971 San Fernando earthquake that strong and ductile steel pipelines withstood ground shaking but were unable to resist the large permanent ground deformation generated by faulting and ground failures.

During the 1971 San Fernando earthquake the steel pipeline system resisted significant seismic forces and the natural gas piping system failed primarily at or adjacent to locations where there were sharp vertical or lateral dislocations or ground ruptures (ASCE, 1984). The pipe was torn or twisted apart at these locations, and the breaks were in the body of pipe, at fittings or at welds, whichever existed at these ground displacement points. At locations where there was severe ground displacement but no ground ruptures, the piping yielded but did not break. O'Rourke and Tawfik, 1983, reported on the effects of lateral spreading on buried pipelines during the San Fernando earthquake. Lateral spreading led to severe damage in this earthquake (EERI, 1986, O'Rourke and Trautmann, 1981, O'Rourke, 1988). Eleven transmission pipelines were affected by lateral spreading and liquefaction-induced landslides. Five pipelines were damaged substantially. The most severe damage occurred in gas transmission pipelines that were deformed by lateral spreading along San Fernando Road, where differential lateral movements as large as 1.7m (5.6ft) were observed. Ground movements due to seismic liquefaction can be extremely large and of great detriment to pipeline safety (Darragh, 1983).

With regard to pipeline failures in liquefied areas, during the 1964 Niigata earthquake the average failure ratio for pipes 100-300mm (4 to 12in) diameter was about 0.97 per km (Katayama et al., 1975). Failure types were reported to be pipe and weld breaks, and joint separations.

3.1.1.3   Secondary Effects

Pipelines have been damaged or destroyed at particular locations due to secondary effects of earthquakes. These secondary effects include flooding caused by failure of water conduits, reservoirs, and dams; hazards from fallen power lines; and explosion hazards when oil tanks and gas lines are ruptured. Experience from past earthquakes indicates that bridges and other supporting facilities can have a significant effect on the performance of oil and gas pipeline systems. An example of secondary damage to pipelines is a pipeline mounted on a bridge that was totally destroyed during the 1976 Tangshan, China earthquake. The bridge length was

12

about 800m (2625ft) and the earthquake intensity was reported to be 9 at this location (Guan-Qing, 1980). Some secondary hazards may be mitigated most effectively through proper siting practices and design of components (Ward, 1990). Design criteria for oil and gas pipeline systems should account directly or indirectly for failure of facilities which may affect the performance of the oil and gas pipeline systems (ASCE, 1983). It is noted that pipeline river crossings can be accomplished by directionally controlled horizontal drilling techniques (Hair and Hair, 1988) which provide an opportunity to avoid deposits susceptible to liquefaction and ground instability by selection of launching and receiving areas and appropriate drilling depths.

3.1.2   Factors Affecting Performance of Pipelines

During past earthquakes the performance of large-diameter oil and gas continuously-welded ductile steel pipelines, with modern quality welds and corrosion protection, has been for the most part satisfactory (Nyman and Kennedy, 1987, ASCE, 1984, ASCE, 1983, EERI, 1986, NOAA, 1973, O'Rourke and Ayala, 1990 ,O'Rourke, 1989). Also, no damage was reported to an apparently well constructed 500-mm (20-in) diameter steel transmission line during the 1989 Armenia earthquake (Schiff, 1989). Further evidence that modern buried welded ductile steel pipelines, which are properly designed, manufactured and installed, generally have performed satisfactorily and have not been ruptured by ground shaking is given by Ford, 1988, Wang, 1988, Lee and Ariman, 1985 , Anderson, 1985, Ariman, 1983, Singhal, 1983, and, Kennedy, 1979. Wave propagation damage to a modern welded steel pipeline is unusual (O'Rourke and Ayala, 1990, O'Rourke, 1988). Many large diameter oil and gas transmission pipelines located in seismic regions have gone through moderately large earthquakes, and their performance has been generally satisfactory.

Modern pipelines are made of ductile steel with full penetration welds, resulting in a system with substantial, inherent ductility (Nyman and Kennedy, 1987, Kennedy et al., 1979). Continuous piping systems must rely upon elastic-plastic properties of the pipe materials to allow enough yielding to prevent rupture or failure during earth movement (Ford, 1983). Because of this ductile behavior, it is expected that buried pipelines generally can withstand considerable soil distortion or differential displacement in cohesive or granular soils without rupture. It is well recognized that toughness (strength and ductility) and flexibility of both pipes and joints are the two governing factors related to the seismic performance of buried pipes (Kubo, 1979). On the basis of damage to gas transmission lines from the San Fernando earthquake, Ariman, 1977, concluded that ductility is the most important factor for seismic design of underground piping systems.

Extensive damage occurred to underground welded-steel transmission pipelines during the 1971 San Fernando earthquake (NOAA, 1973). The most serious damage occurred to an oxy-acetylene-welded pipeline installed about 1930. In the same general area of the San Fernando Valley that experienced extensive ground failures, several newer pipelines installed after 1960 did not experience failure. Before the early 1930s, steel pipelines in California were often constructed under quality control less stringent than that imposed today (EERI, 1986). The newer pipelines were characterized by higher yield strengths (x-grade) and modern arc welding (Eguchi, 1987).

Damage to welded steel pipelines during the 1952 Kern County earthquakes was reported to be more extensive with oxy-acetylene-welded lines than those with electric arc welds. The apparently higher incidence of earthquake damage for oxy-acetylene welds may be related to weld quality (EERI, 1986, McCaffrey and O'Rourke, 1983). Most of the damage to gas lines during the 1971 San Fernando earthquake was caused by tensile failures across oxy-acetylene welded joints. It was unlikely that these failures were related to the type of weld, but rather to the quality of the welds (McCaffrey and O'Rourke, 1983). The quality of the welds is one of the most important factors affecting the earthquake performance of pipelines.

It was reported that damage to gas transmission lines resulting from the 1971 San Fernando earthquake was concentrated in four pre-1931 lines that ranged from 300 to 660 mm (12 to 26 in) in diameter (Nyman, 1987, Johnson, 1983, NOAA, 1973). Most of the breaks were at the welds, but a number occurred between welds (EERI, 1986, Johnson, 1983, McNorgan, 1973, Bagwell, 1973).

Another example of increased number of breaks of oxy-acetylene-welded steel pipes compared to arc-welded steel pipes occurred during the 1964 Niigata earthquake. The average number of breaks of the oxyacetylene-welded steel pipes was five times greater than that experienced by normal arc-welded steel pipe (Eguchi, 1987).

The joints of cast-iron pipe have also been susceptible to damage by earthquake. In the 1923 Kanto earthquake over 4000 pipeline breaks were reported in the Tokyo region. Most of the damage occurred at the joints of small diameter cast-iron pipe which were pulled apart by the earthquake (Eguchi, 1987). Evidence from two major earthquakes in China (1975 Haicheng and 1976 Tangshan) indicate that pipe joints or pipeline portions near them were easily broken, either pulled out, crushed, bent, or sheared into two or many parts, while flexible joints were seldom damaged (Fu-Lu, 1983). For segmented pipelines the adoption of flexible joints with a rubber ring is the best way to reduce the damage (Shoaping, 1983).

Corrosion of pipes and pipelines affects their service life and reduces their ability to resist seismic forces (Ogawa, 1983). Isenberg, 1979, reported that more than half of the leaks in water pipelines attributed to the 1971 San Fernando, 1969 Santa Rosa, and 1965 Seattle earthquakes were in pipes weakened by corrosion. Experience in the petroleum industry indicates that steel pipe can be protected against external corrosion by a combination of coatings and cathodic protection (Hair and Hair, 1988). Corrosion of pipes and welded steel pipelines and methods for protecting them from corrosion are discussed by Isenberg, 1979, and O'Rourke et al., 1985. Pressure surge in pipelines due to seismic excitation may also increase the possibility of failure of pipelines weakened by corrosion (Young and Pardon, 1983, Ogawa, 1983).

3.1.3   Failure Mechanisms

The principal modes of failure for continuous, welded pipelines are direct tensile rupture, beam or local buckling, and excessive bending. For jointed or segmented pipelines, the principal modes of failure include rupture or excessive deformation of individual pipe segments, pull-out or compressive battering of joints, and excessive rotation of joints. There has been substantial research regarding the modes of pipeline failure (e.g., ASCE, 1984; O'Rourke, el al., 1985; O'Rourke, 1988), and all failure modes have been observed in previous earthquakes

14

(e.g., Wang and O'Rourke, 1978; ASCE, 1984 Hall and O'Rourke, 1991), particularly in areas subjected to large permanent ground deformation.

The tensile capacity of a segmented pipeline is generally controlled by the tensile strength or operable pull-out displacement that characterizes the joints. For girth welded steel pipelines, tensile capacity depends on the amount of axial elongation which can be sustained until pipe rupture. Many modern steels have good post-yield characteristics in axial tension. Kennedy, et al. (1977) recommended a maximum strain of one-third the ultimate tensile capacity of pipeline steel for the combined action of axial and bending deformation. Typically, the direct tensile strain capacity ranges from 2 to 5 % for X-grade steels (ASCE, 1984).

Pipelines oriented to sustain tensile elongation in response to permanent ground deformation are able to accommodate relatively large ground displacement by virtue of the ductility of the pipeline steel. This principle is an underlying factor in the recommended design practices for pipelines crossing active faults (ASCE, 1984). Pipelines oriented to accommodate ground movement by means of combined axial tension and bending have performed well under liquefaction-induced lateral spreading. During the 1971 San Fernando earthquake, four large gas and liquid fuel pipelines located on the western side of the Upper Van Norman Reservoir were able to sustain approximately 2.5m (8.2ft) of lateral soil displacement which was directed perpendicular to the longitudinal axes of the lines (O'Rourke, et al., 1990).

Large compressive ground deformation can result in beam buckling, when the pipeline lifts out of the ground, or in local buckling or shell wrinkling, characterized by crippling and distortion of the pipe wall. Experimental results have shown that local wrinkling will begin at strains approximately 15 to 20% of the wall thickness to radius ratio of the pipe (ASCE, 1984). Strains on the order of 4 to 6 times as great generally can be sustained without tearing at a compressive wrinkle (Hall and O'Rourke, 1991). The opportunity for beam buckling is closely related to the depth of cover. For pipeline burial exceeding 0.5 to 1.0m (1.6 to 3.2ft), beam buckling generally will not occur so that only local wrinkling will result under these conditions from excessive compressive strains (Meyersohn and O'Rourke, 1991).

The most probable causes of large compressive strains in buried pipelines that can lead to buckling are fault movement (including creep), landslides and other massive ground movements. Experience has shown that pipelines with bends, elbows, and local eccentricities will concentrate deformation at these features, especially if ground movements develop compressive strains (EERI, 1986). It was observed in a number of sections of 406-mm (16-in) pipe that buckling under compressive forces at fault crossings occurred during the 1971 San Fernando earthquake (Ariman, 1983, 1984). Ring type buckling occurred to a 529-mm (21-in) diameter oil pipeline during the 1976 Tangshan earthquake. Its diameter was reduced by 40 percent. It should be noted that pipeline performance during the 1971 San Fernando earthquake indicated that local compressive forces can be imposed by reverse faulting despite a favorable orientation of the pipeline (McCaffrey and O'Rourke,1983).

The 1971 San Fernando earthquake caused significant damage to underground gas distribution pipelines. Most of these failures occurred at the welds of welded-steel pipelines with gas-welded joints. Pipeline ruptures at welds made before 1930 led to explosions which

15

left craterlike depressions in residential streets (EERI, 1986). After the 1989 Loma Prieta earthquake, gas distribution system failures in the Bay and epicentral areas showed a significant potential for fires. These failures were predominantly in areas of unstable soils (NIST, 1990).

Seismic failure modes of buried pipelines in liquefiable zones were found to be pull-out, breaking, buckling, and crushing (Yeh and Wang, 1985). A state-of-the-art review of the behavior and damage of buried pipelines due to seismic excitation is presented by Mashaly and Datta, 1983.

3.1.4　　　　Remedial Measures

Modern pipeline steels generally can accommodate average tensile strains on the order of 2 to 5 percent without rupture, with local strains of 15 percent or more. In Japan, steel pipes are allowed to have a design strain of 0.3% for a rare earthquake (Singhal, 1983-III). A reasonable criterion, suggested by Hall and Newmark, 1978, for permissible deformation to avoid rupture appears to be in the order of 1 to 2% strain in modern steel pipe at any section. Careful quality control over pipeline manufacture and welding is a necessity for achieving the desired performance under these strains (Nyman and Kennedy, 1987).

Damage to pipelines may be minimized provided that a correct choice of pipe material, type of joints, arrangement of the network, length of segments, location and details of fittings and accessories are made, and as long as pipelines are not located in the vicinity of fault or landslide zones (Fu-Lu, 1983).

Past earthquake fires in Japan have led to the installation of peak acceleration detectors at various locations in gas transmission lines (Schiff et al., 1984). Detection of certain levels of peak acceleration will reduce the pressure in the lines and isolate LPG storage tanks. Higher detected levels of acceleration will stop gas generation and valves to the transmission line are closed so that the system is sectionalized. Very high accelerations will result in purging of gas transmission lines. It is noted, however, that in 1973 McNorgan stated that the installation of earthquake, vibration, or automatic shut-off valves is not a panacea for such situations; such valves could cause severe problems from an operational standpoint.

3.1.5　Summary

With respect to buried gas and liquid fuel pipeline systems, the following lessons learned were summarized by the Earthquake Engineering Research Institute (EERI, 1986):

- Large permanent ground movements are the most severe earthquake hazard affecting gas and liquid fuel lifelines.

- Locations most vulnerable to earthquake damage are pipeline bends, elbows, tees, and local eccentricities, especially if compressive strains develop as a result of permanent ground movement.

- Pipelines made of steel with quality welds and protected against corrosion have performed well during earthquakes even when subjected to permanent differential ground movements.

- Some steel pipelines constructed before or during the 1930's are susceptible to earthquake damage because of relatively weak welds.

3.2 Tanks

3.2.1 Overall Performance Record

During previous earthquakes, many tanks have been damaged by strong ground shaking and some have failed with serious consequences. Because of the wide use of tanks and their vulnerability to earthquakes, many incidents of damage to tanks have been reported (EERI, 1986). During the 1964 Niigata earth quake, oil from ruptured tanks caught fire, damaging two refineries (ASCE, 1984). Waterways were polluted because of oil storage tank failure during the 1978 Miyagi-ken-oki earthquake (ASCE, 1984). Tank damage after the 1989 Loma Prieta earthquake was observed in the epicentral area, and as far as 120km (78mi.) from the epicenter (EERI, 1990). Because of numerous earthquake-induced failures coupled with the potential for fire, pollution, and contamination of surrounding areas, the seismic behavior of liquid storage tanks is a matter of great concern.

3.2.2 Earthquake Effects Causing Failures of Tanks

The predominant hazards for tank farms are ground shaking and liquefaction (Kennedy et al.,1979). Generally tank farms can be sited to avoid or minimize the potential damage associated with fault movement. In the 1964 Alaska earthquake, considerable damage to oil storage tanks occurred over a wide area of Alaska (Eguchi, 1987, ASCE, 1987). Much of the damage was due to the effects of tsunamis, earth settlement, and liquefaction. Eguchi, 1987, reported that experiences from this earthquake have led to significant changes in the design of above-ground storage tanks to resist earthquake forces.

3.2.3 Factors Affecting Tank Performance

Haroun, 1983, reported that tanks with large liquid depth-to-radius ratios frequently suffered structural damage, while shell damage was less common in large capacity tanks which have a large radius and a small depth-to-radius ratio. Overturning moments appear to have been of critical importance in tanks damaged during earthquakes. Seismic excitations produce hydrodynamic pressure at the liquid-shell interface resulting in a lateral force and overturning moment at the base of the tank. There were many reports of tank damage resulting from the 1989 Loma Prieta earthquake. Much of the damage was at soft-soil sites, and it was typically to nearly full tanks. Unanchored tanks with height to diameter ratio exceeding 0.5 were especially vulnerable (EERI, 1990).

17

3.2.4 Failure Mechanisms

The most characteristic type of liquid storage tank damage is a circumferential "elephant's foot" bulge that can form near the base of the tank due to excessive compressive loads in the tank wall (Nyman and Kennedy, 1987,EERI, 1986, Haroun, 1983). Excessive sloshing of tank contents has often resulted in damage to floating and fixed roofs, and tank settling, sliding, or rocking has caused breakage or pull-out at piping connections. Differential settlements of the foundation have also led to tank failure.

In the 1989 Loma Prieta earthquake, damage to unanchored tanks was associated with uplift of the tank walls. Uplift displacements between the shells and foundations of some fully loaded tanks was judged to be between 150 and 200mm (6 and 8in). Failure types included elephant's foot buckle, vertical splits in tank walls, ruptures of elephant's foot buckles, puncture of tanks by restrained pipe, and damage to restrained piping anchored to both tank and foundation (EERI, 1990). Similar, partially filled or empty tanks, adjacent to the damaged ones, were undamaged. During the 1952 Kern County earthquakes, oil-storage tanks were occasionally damaged near their tops by sloshing oil. Floating tops suffered more damage than fixed roofs (ASCE, 1987).

A major oil refinery with about 90 storage tanks in the area of the 1978 Sendai earthquake had three large tanks fail and three others damaged without failure. Bolts around the circumference of a large welded steel-plate water tank were pulled out of their concrete embedment from 25 to 150 mm (1 to 6 in). There was no damage to the base of the tank. Several LPG tanks at the refinery suffered only minor cracks in concrete supports. These tanks were heavily braced with diagonal braces having circular cross sections (EERI, 1986).

Only one of six unanchored ground-based tanks which were 99 percent full was damaged by the 1975 Imperial Valley earthquake. The largest one was damaged and spilled oil. Failure of the fixed steel-plate roof and separation of the perimeter weld around the roof allowed some of the sloshing oil to run down the exterior of the tanks. Four of 18 gasoline and diesel tanks at a tank farm suffered damage in the form of a moderate elephant's-foot bulge. There was no apparent leakage. The tanks were located on concrete ring walls or compacted gravel fill, none were anchored, and most had floating roofs. Compression buckles were more prominent in tanks supported on concrete ring walls than those on gravel fill (EERI, 1986).

Of the 120 vertical unanchored tanks at a refinery, 12 were damaged during the 1985 Chile earthquake. The tanks had capacities between 2,500 and 125,000 barrels. Most tanks appeared to have failed either at the base plate or at the weld between the wall and base plate. Several elephant's foot buckles were observed, and at least four tanks had roof damage when their contents emptied faster than relief valves equilibrated the pressure. Damaged tanks were either full or nearly full at the time of the earthquake. Many of the tanks appeared to have rocked, and differential settlement damaged the pipes exiting some of the tanks at their bases (EERI, 1986).

Many tanks were affected by the 1971 San Fernando earthquake. Damage to two water tanks was reported (EERI, 1986). One of them was a large tank about three-fourths full which showed signs of having rocked on its foundation. Some of the anchor bolts failed in tension and others apparently failed in bond and were pulled up out of their anchorage from

50mm (2in) on one side to 356mm (14in) on the other side. The other tank sustained an outward bulge close to ground level almost all the way around the circumference (elephants-foot buckle). The bulge covered a height of about 508mm (20in) and an amplitude of about 200mm (8in). The outlet pipe and connection broke, the floor plate broke from the walls at one place, and water emptied.

Damage to elevated tanks was reported to fall in the following categories: damage to the support structures, such as stretching of ties, buckling of struts, tearing, warping, and rupture of gusset plates at end connections; separation of clevises, rivets, and bolts; damage to piping and other appurtenances connected to tanks due to tank movement; damage to anchor bolts; damage to the foundation system, which was aggravated in some instances by liquefaction and slope failures (EERI, 1986).

Two elevated tanks were reported to have received minor to moderate damage during the 1979 Imperial Valley earthquake, and a 380,000L (100,000 gal) water tank collapsed. This tank was estimated to be 30m (100 ft) high by 9m (30 ft) at the base and it had four tubular legs braced with tiers of diagonal rods (EERI, 1986).

3.2.5   Design Methodologies

Manos and Clough, 1983, state that there is a need for a realistic prediction of the uplift mechanism and the out-of-round distortional response to be incorporated in the design of tanks, and that foundation flexibility should be considered in the design of free and fixed base tanks. Combra, 1983, notes that there is a need for a new theory concerning tank resistance to lateral force. Haroun and Tayel, 1983, state that with few exceptions, current seismic design codes for ground-base cylindrical tanks neglect the effect of vertical ground accelerations. Research and analyses of storage tanks with regard to their response to earthquakes are reported by Shibata et al., 1983.

3.2.6   Lessons Learned

The Earthquake Engineering Research Institute concluded that the following lessons can be learned from the performance of gas and oil storage tanks in past earthquakes (EERI, 1986):

- Much of the poor earthquake performance of tanks can be attributed to the fact that unpressurized tanks, though structurally very efficient for vertical gravity loads, are not structurally efficient for lateral earthquake forces.

- The performance of anchor bolts at tank bases and towers has been poor in many cases. Anchor bolt failure during many seismic events shows that more thought must be given to their use and detailing, and to whether bolts are needed.

- Enough freeboard must be provided to prevent waves from contacting the roof system. In many cases, insufficient freeboard has led to damage of roofs by sloshing liquid.

19

- Flexible joints or adequate piping flexibility that allows for expected relative motion between tanks and piping should be used. The failure of rigidly attached piping often causes tank contents to be released.

- Because damage to elevated tanks affects the supporting structures while the vessel remains intact, better design of the struts and detailing of the connections is required.

3.3   Structures and Above Ground Support Facilities

3.3.1   Overall Performance Record

Control systems and communications are critical for safe and continuous operation of gas and liquid fuel pipeline systems and are vital for emergency response. Observations after the 1987 Ecuador earthquakes (Crespo, O'Rourke, and Nyman, 1988) suggest that these facilities need more attention. At the Salado Pump Station in Ecuador, control panels were damaged, the main control valve was buried and jammed in an open position by landslide debris, electrical power and auxiliary generators were out, and the radio communications antenna had buckled and become inoperable. At most facilities, control and communication systems have been procured and installed without regard to earthquake resistance, and outside California, anchorage of these critical items has often been inadequate or nonexistent (Nyman, 1991).

Many of the support facilities are similar to other industrial facilities. The major difficulty in evaluating seismic performance of industrial facilities results from their diverse geographical locations, special design considerations, different dates of construction, and from the fact that criteria for seismic design vary from structure to structure (EERI, 1986). It is well known, that port and harbor facilities, including piers, docks, quays, and landings, are particularly susceptible to the effects of strong earthquakes (EERI, 1986).

Experience indicates that modern facilities designed and constructed in accordance with modern United States seismic practice, with particular attention given to adequate anchorage of equipment, can be expected to sustain no significant loss of operating function when subjected to high-level resonant ground motion (ASCE, 1984). The components of oil and gas pipeline systems which satisfy modern seismic design criteria, have in general exhibited good behavior in past earthquakes. This includes the above ground components such as compressor stations, pumping stations, and control stations (ASCE, 1984, EERI, 1986). Proper anchorage of equipment, including items in the control center, can greatly reduce damage and minimize injury to personnel (EERI, 1986).

Four major modern industrial facilities subjected to severe ground motions during the 1985 Chile earthquake performed well, although minor damage was sustained, none of these facilities were shut down (EERI, 1986). Industrial facilities in general were also not seriously damaged by the 1989 Loma Prieta earthquake.

Limited information is available pertaining to pipeline stations and terminal facilities, however there is considerable information available on the performance of similar facilities which include refineries and power plants. Experience has shown that the seismic performance of large equipment and machinery is vital and the most important design consideration is providing adequate anchorage (Nyman and Kennedy, 1987, ASCE, 1984, ASCE, 1974, EERI, 1986, Anderson, 1985).

It has been found from pipeline projects that critical electrical equipment and instrumentation which includes computers, valves, motors, control panels, and pressure switches exhibit good resistance to seismic shaking when securely anchored. Lack of anchorage or inadequate anchorage of equipment has led to rupture of electrical connections and thus failure of electrical power supply (Nyman, 1987, ASCE, 1984, EERI, 1986).

Bettinger, 1980, reported that limited experience with gas compressor stations in earthquakes has been favorable. Gas compressor stations are conservatively designed and built, and further provisions for seismic resistance do not appear to be warranted.

Schiff and Yanev, 1989, reported on damage caused by the 1989 Armenian earthquake to two non-nuclear power facilities. They noted that equipment anchorage could be improved.

Underground facilities such as vaults and manholes were not damaged structurally as a result of the 1971 San Fernando earthquake. This was true for cast in place and prefabricated vaults (ASCE, 1974).

3.3.2 Design Methodologies

Current design practices recognize that equipment outages due to earthquakes can best be mitigated by proper design provisions to prevent sliding or tipping of equipment and falling debris (Anderson, 1985). Mechanical and electrical equipment and instrumentation serving pipeline transmission systems can be vital for maintenance of a sufficient level of service and for control and emergency procedures in the event that damage occurs.

For critical equipment and instrumentation, a seismic qualification procedure may be implemented to demonstrate the capability for continued or uninterrupted operation (Nyman, 1987, Anderson, 1985, Anderson and Nyman, 1979, 1977). Seismic qualification has been of great importance for many years in nuclear power plants. A useful guide for seismic qualification can be gained from examining the performance of equipment and instrumentation during previous destructive earthquakes.

EERI, 1986 reported that the state of practice of earthquake resistant design of industrial buildings needs to be improved in some areas. The level of damage has been significant, even to modern structures, during some recent moderate earthquakes. Foundation performance contributed to some of the problems, but most of the damage was a direct result of poor connections and inadequate anchorage.

3.3.3 Lessons Learned

The Earthquake Engineering Research Institute concluded that the following lessons can be learned from the performance of gas and oil storage tanks in past earthquakes (EERI, 1986):

- The information available suggests that above-ground facilities that are designed to resist seismic effects suffer limited damage in earthquakes.

- Proper anchorage is important in preventing damage to mechanical equipment. When outages occur as a result of an earthquake, the cause can usually be traced to falling debris, collision with other items, sliding (with subsequent rupture of electrical connections or piping), or failure of the electrical supply.

4. AVAILABLE DESIGN CRITERIA, REMEDIAL MEASURES, STANDARDS, AND DESIGN GUIDES

4.1 Introduction

This section addresses the design of new systems, as well as the retrofitting of existing systems. Three levels of sophistication are identified. State of the art methodology is at the highest level. At this level, engineers can design a pipeline or storage tank using the latest methodologies in site exploration and mathematical modeling, reflecting our present state of knowledge. For very large and important projects, such as the Trans-Alaskan Pipeline, this has been done, and the only question that arises is whether the present state of knowledge is adequate. At the second level are established design criteria and methodologies. At the third level are standards, codes and design provisions which can be made mandatory and thereby establish a minimum level of performance. It is the third level which is of greatest interest in this report, because it is important to determine whether existing standards, codes or design provisions, when minimally complied with, will produce systems which will perform adequately. However, it is also important to establish whether the present state of knowledge is adequate.

In addition to a discussion of the topics, relevant information from the technical literature is presented. The views expressed in this latter information, which is typed in italics, are not necessarily consistent with those expressed by the authors of this report.

4.2 Design Criteria

4.2.1 Development of Design Criteria

The development of seismic design criteria first became of real interest to the petroleum industry following the damage to oil storage facilities during the 1933 Long Beach, California earthquake. Development of seismic design criteria for critical facilities occurred relatively slowly until about 1960 when the advent of nuclear power plants triggered the need for developing and employing modern earthquake engineering principles and practices.

Following the 1971 San Fernando earthquake, interest in the effects of earthquakes on lifeline systems appreciably increased. In 1974 the American Society of Civil Engineers formed the Technical Council on Lifeline Earthquake Engineering (TCLEE). Since that time there has been an increasing number of technical papers on the subject.

Seismic design procedures for gas and liquid fuel pipelines were proposed by Kennedy et al., 1977, Hall and Newmark, 1977 and Hall and Kennedy, 1980. By 1984 the Gas and Liquid Fuel Lifelines Committee of TCLEE developed "Guidelines for the Seismic Design of Oil and Gas Pipeline Systems" (ASCE, 1984).

The ASCE "Guidelines for the Seismic Design of Oil and Gas Pipeline Systems" (1984) are intended primarily for engineers engaged in the design of most major components of gas and liquid fuel pipeline systems. The document also provides guidance to pipeline company management, disaster recovery agencies, regulatory agencies, and insurance groups. The document provides general guidance on design, construction, operation, maintenance, and upgrading of systems and components common to pipeline systems.

Hall, 1987, reported that considerable additional work can be done to reduce damage to pipelines and facilities when subjected to moderately severe earthquakes. Improvements in earthquake engineering center around sound engineering practice that can be attained only if there is a good understanding of the expected behavior of pipelines and related facilities.

4.2.2 Current Design Criteria

4.2.2.1 Pipelines

As noted in Section 3, modern continuously-welded ductile steel pipelines performed well in past earthquakes. However, even with modern ductile pipelines, there are problems in areas of severe soil deformations, at connections to structures, at locations of bends and junctions, and at locations where pipelines are threatened by failures of bridges, dams, and electrical systems, or by earthquake related phenomena such as landslides, tsunamis, seiches, and soil liquefaction.

Thus there is a need for seismic provisions which promote good construction and quality control. The most important aspect of such provisions should be proper siting, designed to avoid hazardous conditions. It is important to recognize that some damage under severe conditions should be anticipated even in pipelines which were designed to be earthquake resistant. Thus, there is also a need for monitoring and emergency shutdown systems which will minimize the environmental and economic consequences of pipeline failures. Analytical, laboratory, and field studies of performance of gas and liquid fuel pipelines should be conducted to develop improved design, assessment and retrofit practices.

Important aspects of pipeline design as discussed in the current technical literature are presented below:

Nationally applicable design and construction provisions for new lifelines, and strengthening provisions for existing lifelines should be developed. In contrast to buildings, and except for highway structures, no nationally applicable design and construction practices are available for new and existing lifelines (NIST, 1990). Eguchi, 1987, reported that with the exception of the Trans-Alaskan Pipeline, very little has been done in the area of system and component performance criteria development for oil and natural gas pipeline systems. The primary reason for the lack of criteria is that requirements tend to differ from one system to another, thus influencing the level of performance.

Many studies, analyses and recommendations were made following the 1971 San Fernando earthquake. It was concluded that there does not appear to be an economical method for fully preventing damage to underground lines due to extreme differential ground movements (Nyman, 1987,Johnson, 1983). Eguchi,1987, reported that the majority of oil and natural gas pipeline system components in the United States are highly vulnerable to earthquakes either because newer seismic design procedures have not been applied at the appropriate level or because of the severity of hazards in the areas they occupy have not been adequately assessed.

The seismic design criteria formulated for each of the pipeline system components should provide estimates, and the basis for such estimates of forces, ground movements or other vibratory motion effects that would be expected for a postulated design earthquake (ASCE, 1984). Most pipeline systems, especially in regions of high seismic exposure, are subject to restrictions and performance requirements of federal and state governmental regulatory agencies as well as those of the facility owner (ASCE, 1984). The pipeline as well as pumps, compressors, flow monitoring and control equipment and other parts of a facility which are critical for continued operation and system control normally should be designed to service a major earthquake with almost no damage. However, structures housing this equipment, storage buildings and other structures not directly affecting the operation of a pipeline facility could experience large inelastic deformations provided possible interaction between the structure and enclosed components does not compromise safety or the operation of critical components of the facility ASCE, 1984).

Pipeline steels generally can accommodate average tensile strains on the order of 2 to 5 percent without rupture, with local strains of 15 percent or more. In Japan, steel pipes are allowed to have a design strain of 0.3% for a rare earthquake (Singhal, 1983-III). A reasonable criterion, suggested by Hall and Newmark, 1978, for permissible deformation to avoid rupture appears to be in the order of 1% to 2% strain in modern steel pipe at any section. Careful quality control over pipeline manufacture and welding is required for achieving the desired performance under these strains (Nyman and Kennedy, 1987).

For optimal design of gas transmission networks, the following design variables need to be determined: number of compressor stations, compressor station locations, lengths of pipeline segments between compressor stations, diameters of pipeline segments, and suction and discharge pressures at each compressor station (Edgar et al., 1978). Damage to pipelines may be minimized provided that a correct choice of pipe material, type of joints, arrangement of the network, length of segments, location and details of fittings and accessories are made, and as long as pipelines are not located in the vicinity of fault or landslide zones (Fu-Lu, 1983).

Under extreme earthquake conditions, it is reasonable to permit larger movements for pipelines, except at restraints such as anchors, valves or pump stations, than for plant facilities (Darragh, 1983). Experimental studies have been conducted on the mechanical behavior of PVC pipelines subjected to ground subsidence (Takada, 1983).

4.2.2.2   Storage Tanks

Many storage tank failures have been caused by earthquakes. Thus, the need for special seismic design provisions is generally recognized. Such provisions require consideration of anticipated lateral and vertical forces. Criteria for estimating anticipated earthquake effects have been developed (Wozniak and Mitchell, 1978), and were incorporated in existing standards (API 650, AWWA D 100). In addition to consideration of inertial forces, design criteria should also deal with preferential tank geometry (tanks with large fluid depth to diameter ratio are particularly vulnerable), siting, and secondary containment to minimize the effects of potential spills. Defensive siting and secondary containment are required for LNG tanks, but not specifically mentioned in design provisions for other fuel storage tanks. As in the case of pipelines, the most important aspect of seismic design is defensive siting, which avoids potential liquefaction hazards and sites where the ground motion is amplified, as well as sites subjected to tsunamis, seiches, other types of flooding, landslides and ground rupture. Unlike transmission pipelines and distribution piping networks, storage tanks in most instances can be located to avoid special site hazards.

Design criteria for above-ground tanks are provided in ASCE, 1984, API 650, AWWA D 100, and for buried tanks in Army Manual TM 5-809-10-1, "Seismic Design Guidelines for Essential Buildings". Haroun, 1990, discusses the API 650 and AWWA D 100 design procedures. The seismic loads in these two most commonly used standards for tank design are based on a mechanical model derived by Housner, 1957, for rigid tanks. Recent versions of the standards have adopted an increase in the acceleration coefficient, which represents the short-period amplified acceleration due to shell deformation. This value of acceleration coefficient, in general, is specified independently of the tank dimensions and support condition. The lateral base shear force is determined from a number of coefficients for site location, natural period and soil profile. For such computations, the input requirements consist of a zone coefficient, a site factor, the response period, and the effective masses and their elevations.

The use of a response spectrum is encouraged by the AWWA standard for sites that might experience severe ground motion during the life of the structure. When the response spectrum is selected, the accelerations obtained from the spectrum are substituted for the seismic coefficients. The API 650 AWWA D 100 standards use somewhat different methods for determining seismic coefficients. In the API standard, the specified seismic coefficient is multiplied by an importance factor to obtain an effective seismic coefficient, whereas in the AWWA standard, the effective seismic coefficient is determined by multiplying a set coefficient by a "structure coefficient" which is different for anchored and unanchored tanks. The computation of forces due to convective motion of fluid is also slightly different in the two standards.

The bending moment at the shell base is used for evaluating the compressive and tensile forces in the tank shell. The allowable earthquake compressive stress consists of the static allowable stress plus a stabilizing stress due to the internal liquid pressure with the sum increased by a specified amount. The stabilizing stress depends on geometric terms and a pressure stabilizing coefficient. Overturning moments, including those arising from the pressure variations on the base, are computed for the design of the foundation.

Other important aspects of storage tank design as discussed in the current literature are presented below:

Manos and Clough, 1983, state that there is a need for a realistic prediction of the uplift mechanism and the out-of-round distortional response to be incorporated in the design of tanks, and Combra, 1983, notes that there is a need for a new theory concerning tank resistance to lateral force. Foundation flexibility is another factor that should be considered in the design of free and fixed base tanks (Manos and Clough, 1983). Research and analyses of storage tanks with regard to their response to earthquakes are reported by Shibata et al., 1983. With few exceptions, current seismic design codes for ground-based cylindrical tanks neglect the effect of vertical ground acceleration (Haroun and Tayel, 1983).A better understanding of the behavior of unanchored tanks, an assessment of the effect of the vertical component of ground motion on tank response, and an acceptable estimation of shell strength against buckling are needed (EERI, 1986).

4.2.2.3   Structures and Support Facilities

Design criteria for structures are well defined and reference to various existing standards can be made for design direction and appropriate formulations (e.i., NEHRP Recommended Provisions; BSSC, 1988; SEAOC, 1987; UBC; BOCA). Most regions in the country legally adopted seismic design standards for buildings. However these provisions may not be sufficient for industrial type facilities. Seismic design provisions for support facilities are also available (Anderson, 1985). The most important aspect of these provisions is to provide resistance to tipping, sliding and uplift. The equipment itself, such as pumps and compressors, does not seem to be particularly vulnerable to earthquake shaking. Nevertheless, it has been suggested that a seismic qualification procedure for peripheral equipment, similar to that used for nuclear power plant equipment, could provide the capability for uninterrupted operation of the transmission system.

Important aspects of design practices as discussed in the current technical literature are presented below:

Ground shaking is a major design factor for pump stations and marine terminal facilities, tank farms, above ground sections of pipeline and above ground structures in general. The effects of tsunamis or seiches often must be incorporated in the site selection and design of marine terminal facilities (ASCE, 1984).

Current practices recognize that equipment outages due to earthquakes can best be mitigated by proper design provisions to prevent sliding or tipping of equipment and falling debris (Anderson, 1985).. For critical equipment and instrumentation, a seismic qualification procedure may be implemented to demonstrate the capability for continued or uninterrupted operation (Nyman, 1987, Anderson, 1985, Anderson and Nyman, 1979, 1977). Seismic qualification has been of great importance for many years in nuclear power plants. A useful guide for seismic qualification can be gained from examining the performance of equipment and instrumentation during previous destructive earthquakes.

It was reported by EERI (1986) that the state of practice of earthquake resistant design of industrial buildings needs to be improved. The level of damage has been significant, even to modern structures, during some recent moderate earthquakes. Foundation performance contributed to some of the problems, but most of the damage was a direct result of poor connections and inadequate anchorage (EERI, 1986).

4.3   Emergency Response, Evaluation, Repair and Retrofitting

4.3.1   Emergency Response

The previous section deals with the design of new fuel pipeline systems and their components. However, most of the risk is associated with existing systems which were not designed to be earthquake resistant. For such systems it is important to have contingency plans to deal with various types of anticipated earthquake damage and monitoring systems which will provide information in case of earthquake damage.

26

The major West Coast pipeline companies have in general included earthquake planning in their emergency procedures. There are cooperative agreements among a number of major companies to share their resources in the event of a major emergency such as an earthquake. The emergency plans appear according to Nyman, 1987, to be a model for utilities in other seismic risk areas such as the New Madrid area and the East Coast.

Recommendations for land use measures are of interest. Recent laws in California call for seismic hazards mapping, such as the Alquist Priolo Act (fault zones) and the Seismic Hazards Mapping Act of 1990 (liquefaction, landslides, and site amplification). They are designed to identify zones of increased risk from large permanent and transient movements. Given the co-existence of critical gas and liquid fuel lifelines and statutory zones of ground failure hazards, it can readily be anticipated that land use planning will play an important role in future measures to curb earthquake risk to lifeline systems.

Emergency response practices depend on emergency planning and preparedness. The National Transportation Safety Board (NSTB) has questioned the adequacy of measures taken to protect public safety near pipelines. Their interest is in more effective land use planning and policies, damage prevention, and in more responsive emergency preparedness programs (TRB, 1988). Thus, plans for improvements in seismic resistant practices should be coordinated with current recommendations for enhancing general practices for pipeline safety, such as those presented by the Committee for Pipelines and Public Safety of the Transportation Research Board (TRB, 1988).

Summarized below are observations on these problems taken from the current technical literature:

Seismic risk assessment methods for natural gas and oil pipeline systems discussed by Eguchi, 1987, allow natural gas utilities and oil companies to better understand the weaknesses of their system and thus understand where to concentrate most mitigation or response planning efforts. The results of studies by several major gas utility companies in California are being used to: (1) identify vulnerable pipeline elements, (2) estimate probable service levels after a major earthquake, and (3) test current emergency response plans (Eguchi, 1987). Eguchi, 1987, described a seismic risk methodology for natural gas and oil systems intended to identify weak links within a system, establish minimum performance standards for facilities, assess minimum performance for the system with regard to an earthquake, test various mitigation, retrofit, and/or design strategies, and to test emergency response plans.

Past earthquake fires in Japan have led to installation of peak acceleration detectors at various locations in gas transmission lines (Schiff et al., 1984). Detection of certain levels of peak acceleration will reduce the pressure in the lines and isolate LPG storage tanks. Higher detected levels of acceleration will stop gas generation and valves to the transmission are closed so that the system is sectionalized. Very high accelerations will result in purging of gas transmission lines. It is noted that in 1973 McNorgan stated that the installation of earthquake, vibration, or automatic shut-off valves is not a panacea for such situations, unless an individual is capable of determining the where and when of earthquakes. Installation of such valves could cause more problems from an operational standpoint than they might solve (McNorgan, 1973).

Japan has had experience with post-earthquake operation of gas systems. Many storage facilities required strengthening. Special criteria for their strengthening were developed (Schiff et al., 1984).

4.3.2 Evaluation

To determine whether fuel transmission systems should be protected against earthquake effects by retrofitting, it is important to evaluate their condition and assess their damage potential. Another problem is evaluation after an earthquake has occurred. In the case of buried pipelines this poses a difficult problem because they are not directly accessible for inspection. The problem is compounded by the fact that pipelines may have sustained damage which requires repair even if they continue to function normally.

Hale, 1984, notes that internal pipe inspection can be carried out by (1) visual (closed circuit television or conventional film camera), (2) electromagnetic, and (3) ultrasonic procedures. It is noted that a closed-circuit TV system is available for piping systems down to 1 1/2 in (40-mm) diameter. Electromagnetic devices have been used to inspect more than 100,000 miles (160,000 km) of pipelines and gas distribution lines from 4 to 48-in (100 to 1220-mm) diameter over the last 20 years. Newly developed on-line inspection vehicles can pinpoint flaws to within a few feet, locate significant defects, and make automatic analysis of data without interfering with pipeline operation.

Summarized below are other observations on these problems taken from the current technical literature:

> *Eguchi, 1987, stated that one of the most critical problems faced by oil and natural gas pipeline operators after a major earthquake is the immediate detection and isolation of damage to the system. Currently (1987) no system or methodology exists for early damage detection of lifelines following an earthquake.*

> *Quantitative models for identifying earthquake damages to pipelines have only recently been developed. Those models available still do not satisfy prediction needs entirely (ATC, 1985). A methodology to perform seismic hazard analysis of geographically extensive regions has also been presented (Monzon-Despang and Shah, 1983).*

4.3.3 Repair

Buried pipelines can be severely deformed by ground distortions which may or may not lead to failure. Current practice according to Nyman, 1987, calls for cutting out and replacing sections of pipelines that have experienced deformations and have been determined to be unacceptable for safe operation. An account of repairs to a natural gas distribution system with some information about a transmission pipeline system is given by Johnson, 1983.

4.3.4 Retrofitting

Retrofitting is necessary to update equipment for more efficient and economic operation and in a form to preclude damage in an earthquake (Hall, 1987). Renovation of existing piping involves rebuilding or restructuring so that it will provide many more years of service without removing the buried pipe and replacing it with new piping material (Hale, 1984). Considering the amount of cast iron gas mains still in existence and operation, plus a growing mileage of aging steel mains with corrosion problems, a substantial need exists for the renovation of current pipeline systems.

Hale, 1984, has reported on renovation techniques currently used by distribution companies for improving the performance of aging mains, particularly those composed of cast iron and corroded sections of steel. Some of these renovation techniques also qualify as measures to improve seismic performance by enhancing the strength and continuity of the piping. Renovation techniques, which can improve seismic performance, include external encapsulation of joints and fittings (generally with a molded polyurethane fitting), insert renewals with plastic pipe, and pressure relining of pipelines with plastic pipe.

In situ cleaning and coating of oil and gas pipelines have gained popularity as operators discover that it is much less costly to renovate troublesome lines than to replace them (Hale, 1984).

To minimize damage from earthquakes, either actual or potential, a rigorous continuing inspection and maintenance program is required (ASCE, 1984). Inspection and maintenance programs should, in general, consist of (1) as-built documentation, (2) inspection plans which include any needed measurements and their frequency, and (3) maintenance and repair plans. The as-built documentation is important so that any differences from design or design assumptions are recognized, documented, and evaluated for their effect on seismic performance. The inspection plan should include a recognition of those key components of seismic design which are required to ensure the design integrity and a scheme for monitoring those components. Measurement programs vary from accurate location of reference monuments pertaining to pipeline system routes and components and ground movement detection devices, to installation and reading of strain gages and other measuring instrumentation. Monitoring frequency should be scheduled to permit early detection of changes in field conditions or in the condition of the facilities which could increase the exposure to seismic hazards. The repair plan depends on anomalies uncovered in the inspections and also includes recognition of recurring problems which have been periodically corrected along with standard corrective actions implemented by operating personnel.

Potentially unstable slopes in the vicinity of the pipeline route should be inspected periodically to determine whether changes have occurred in the field which could change drainage patterns. Another important inspection and maintenance activity often overlooked is the identification of vulnerability to non-structural damage caused by the overturning of unanchored equipment, furniture, and storage racks. These unanchored items should not be allowed to move or overturn so that adjacent critical components are damaged or otherwise rendered nonfunctional (ASCE, 1984).

Consideration should be given to the proper storage of repair parts, tools, and equipment so that during an earthquake they are not damaged or cause injury to workers. These items need to be readily available and usable following an earthquake.

Most pipeline systems experience upgrading or replacement during their operational life. Care should be taken to ensure that changes to the facilities and components satisfy the original or updated seismic criteria and specifications (ASCE, 1984).

4.4 Standards, Codes, and Design Guides

4.4.1 Summary of Available Codes and Standards

4.4.1.1 Codes and Standards for Pipelines

Currently used codes and standards for the design of pipelines are listed below:

- Code of Federal Regulations, 49 CFR, Transportation:

    Part 190, "Pipeline Safety Program Procedures"
    Part 191, "Transportation of Natural and Other Gas by Pipeline"
    Part 192, "Transportation of Natural and Other Gas by Pipeline, Minimum Federal Safety Standard"
    Part 195, "Transportation of Hazardous Liquids by Pipeline".

- ASME (ANSI) B31.4, "Liquid Transportation Systems for Hydrocarbons, Liquid Petroleum Gas, Anhydrous Ammonia, and Alcohols".

- ASME (ANSI) B31.8, "Gas Transmission and Distribution Piping Systems".

- ASME Guide for Gas Transmission and Distribution Piping Systems-1983 (ASME, 1983b). (Recommended practices of ASME Gas Piping Technology Committee)

- API Standard 1104, "Welding of Pipelines and Related Facilities".

- California Public Utility Commission (CPUC), General Order 112-D Rules Covering Design, Construction, Testing, Maintenance and Operations of Utility Gas Gathering, Transmission, and Distribution Piping Systems; Liquid Natural Gas Facilities Safety Standard.

- NAVFAC DM - 22, Petroleum Fuel Facilities

- Department of the Army, CEGS-02685, Guide Specification for Military Construction, Gas Distribution Systems.

- Army TM 5-809-10-1, Navy NAVFAC P-355.1, Air Force, AFM 88-3, Chap. 13, Sec.A, Seismic Design Guidelines for Essential Buildings, pg. 7-2, ¶ 7-7, "Buried Structures".

4.4.1.2 Seismic Design Provisions in the Codes and Standards for Pipelines

Federal regulations for natural gas pipelines do not contain explicit requirements for seismic design. Implicit requirements could be read into Subpart C Section 192.103 of Part 192 of 49 CFR, "General", which requires pipes to withstand "anticipated external pressures and loads that will be imposed on the pipe after installation". No mention is made of compliance with anticipated ground strains and deformations. Section 840.1 "General Provisions" of ASME Standard B31.8 mentions "unstable ground" and "earthquake induced stresses" as "conditions that may cause additional stress to the line and its appurtenances and shall be provided for". This is an explicit requirement for consideration of seismic loads where they exist, but not for avoidance of hazardous sites. Again no mention is made of compliance with anticipated ground strains and deformations.

The Federal regulation for transportation of hazardous liquids, Part 195 of 49 CFR, does have an explicit requirement for earthquake resistant design of pipelines in ¶ 195.110, "External Loads", which mentions earthquakes. Again, there is no requirement for compliance with ground strains and displacements. Similarly, ASME Standard B31.4 in Section 401.5.3 requires consideration of the dynamic effect of earthquakes in the design of piping systems in regions where earthquakes are known to occur.

NAVFAC DM 22 for petroleum fuel facilities references ASME Standard B31.4, and CEGS Specification 02685 references ASME Standard B31.8. Otherwise, these specifications do not have an explicit requirement for seismic design. The Seismic Design Guidelines of the Armed Services (TM 5-809-10-1, etc.) mention buried structures, but do not recommend seismic displacements or forces for pipelines.

In summary, consideration of earthquake forces is explicitly required in the federal regulations for liquid fuel pipelines, but it is not explicitly required in the federal regulations for natural gas pipelines. The ASME standards require consideration of earthquake forces for all pipelines. These standards apply by reference to military construction.

It could be argued that the ASME requirements, when interpreted by engineers who are familiar with the earthquake problem, could produce adequate designs. However, the lack of specificity in the provisions, as well as the failure to mention the need for siting studies, raises serious questions about the adequacy of current seismic design provisions.

Quality control of welds is a key factor in the construction of earthquake-resistant pipelines. Current U.S. Government regulations for welding of gas and liquid fuel steel pipelines are given in 49 CFR Part 192 and Part 195. Both Parts 192 and 195 make reference to API 1104, which is frequently used in the gas and liquid fuel industries to establish procedures for weld quality and welder qualifications. The welds described are continuous circumferential welds at straight butt-end connections made by means of electric arc procedures. API Standard 1104 also presents methods for the production of high-quality radiographs through the use of qualified technicians and approved methods and equipment, to ensure the proper analysis of the welding quality. This standard is intended to apply to the welding of piping used in the compression, pumping, and transmission of crude petroleum, petroleum products and fuel gases, and to distribution systems when applicable. Details on welding specifications, standards of acceptability, and radiographic procedures are included in the standard.

4.4.1.3 Codes and Standards for Storage Tanks

Currently used codes and standards for the design of storage tanks are listed below:

- Code of Federal Regulations, 49 CFR, Transportation: Part 193, "Liquefied Natural Gas Facilities; Federal Safety Standards"

- API Standard 650, "Welded Steel Tanks for Oil Storage"

- ANSI/AWWA D100, "Standard for Welded Steel Tanks for Water Storage"

- NFPA 59 A, "Standard for the Production, Storage, and Handling of Liquefied Natural Gas (LNG)".

- California Public Utility Commission (CPUC), General Order 112-D, Rules Covering Design, Construction, Testing, Maintenance and Operations of Utility Gas Gathering, Transmission, and Distribution Piping Systems; Liquid Natural Gas Facilities Safety Standard.

- NAVFAC DM - 22, Petroleum Fuel Facilities

- Army TM 5-809-10-1, Navy NAVFAC P-355.1, Air Force, AFM 88-3, Chap. 13, Sec.A, "Seismic Design Guidelines for Essential Buildings", pg. 7-2, ¶ 7-7, "Buried Structures".

4.4.1.4 Seismic Design Provisions in the Codes and Standards for Storage Tanks

API Standard 650 has detailed seismic design requirements in Appendix E. Similar requirements are incorporated in Section 13 of ANSI/AWWA D100.

Part 193 of 49 CFR specifies site investigation and siting requirements for LNG facilities, which include determination of earthquake forces and earthquake induced displacements in Section 193.2061, "Seismic Investigation and Design Forces". Requirements for subsurface investigation are specified in Section 193.2065, "Soil Characteristics". Site investigations are generally performed in accordance with the recommendations in NBSIR 84-2833 (Kovacs et al., 1984). Another provision of this latter standard, which reduces seismic hazards, is a secondary impoundment dike requirement, provided in Sections 193.2149 through 2165. Similar provisions are included in NFPR 59 A.

Part 195 of 49 CFR requires compliance with API Standard 650 by reference in Appendix A.

The armed forces seismic guidelines (Army TM 5-809-10-1) has provisions for ground-based vertical tanks on page 7-2 and for elevated tanks on page 7-1. These provisions are explicit, but more concise than the ASME provisions. Page 7-2 of the document also addresses buried structures, and could be applied to buried tanks. It recommends seismic design for large-diameter buried structures, but not for pipelines.

It is noted that the API and ANSI/AWWA standards, as well as the armed forces standards, while spelling out detailed seismic design requirements, do not address the siting and secondary impoundment problems. While the Federal provisions, which are intended for LNG storage tanks, may be too restrictive for other liquid fuel and gas storage tanks, siting and secondary storage requirements seem also appropriate for these latter facilities.

4.4.1.5 Codes and Standards for Structures and Support Facilities

The most frequently used codes and standards are listed as follows:

- BSSC, "NEHRP Recommended Provisions"

- ICBO, "Uniform Building Code"

- Southern Building Code Congress Int., "Standard Building Code"

- BOCA, "National Building Code"

- SEAOC, "Recommended Lateral Force Requirements"

- NAVFAC DM - 22, "Petroleum Fuel Facilities"

- Army TM 5-809-10-1, Navy NAVFAC P-355.1, Air Force, AFM 88-3, Chap. 13, Sec.A, "Seismic Design Guidelines for Essential Buildings", pg. 7-2, ¶ 7-7, "Buried Structures".

- ASCE Technical Council on Lifeline Earthquake Engineering, "Advisory Notes on Lifeline Earthquake Engineering".

- ANSI/IEEE Standard 344-1987, "Recommended Practices for Seismic Qualification of Class 1E Equipment for Nuclear Power Stations".

- ANSI\IEEE Standard 693-1984, "IEEE Recommended Practices for Seismic Design of Substations".

4.4.1.6 Seismic Design Provisions in the Codes and Standards for Structures and Support Facilities

Seismic design provisions for structures are incorporated in most building codes in seismic areas and in the model codes listed above. Except for hazardous sites, such as sites in the vicinity of faults and in areas subjected to landslides, liquefaction, and tsunamis, these provisions will generally promote adequate earthquake resistance. Army Manual TM 5-809-10-1 also has adequate provisions for earthquake resistant construction.

Nevertheless, supplemental provisions for the seismic qualification of essential mechanical, electrical, and control equipment components, as well as tiedown requirements for these components should receive consideration. Conservative provisions of this nature were developed for nuclear power plants (ANSI/IEEE 344-1987), and less conservative provisions also were developed for the design of electrical substations (ANSI/IEEE 693-1984).

4.4.2 Comments on Available Codes and Standards

4.4.2.1 Siting

Siting studies for upgrading earthquake-resistant construction practices need to be considered from a perspective of practicality and cost. Gas and liquid fuel transmission lines are large, high-pressure structures, and it is sensible and economically justifiable to scrutinize their vulnerability to natural hazards which can affect their integrity. Seismological studies should be required to locate causative faults, identify design level and maximum credible earthquakes, and provide appropriate attenuation relationships for the estimation of strong ground motion. Geotechnical studies should focus on identifying active faults, areas of potential liquefaction, potential landslide zones, and areas of exceptionally severe site amplification. The potential ground deformations caused by liquefaction, particularly those associated with lateral spreading, should be estimated. Special attention should be focused on river crossings. Consideration should be given to siting and construction techniques (such as directional drilling) to minimize exposure to liquefaction-induced ground failure at river crossings.

In contrast to siting of transmission pipeline systems, siting studies are generally not needed for distribution piping. The vast number of distribution mains and their restriction to street and highway rights-of-way make the concept of siting studies impractical for each pipeline in this type of system. However, special measures, such as ductile piping and welded connections, could be required for new distribution pipes installed in potentially unstable areas which have been mapped. These areas would include fault zones, which in California have been mapped as Alquist-Priolo Special Study Zones, and zones of ground failure hazards which currently are being mapped on a state-wide basis in California. The Seismic Hazards Mapping Act of 1990 enacted by the State of California calls on the State Geologist to undertake a statewide seismic hazard mapping and technical advisory program, in order to assist cities and counties in fulfilling their responsibilities for protecting the public health and safety against the effects of strong ground shaking and other seismic hazards (Tobin, 1991).

Comments in the technical literature:

Many gas and liquid fuel lifeline (GLFL) facilities are sited near the coast; thus, marine as well as terrestrial seismic hazards must be considered in the design. The terrestrial seismic hazards that can affect GLFL facilities are: (1) ground failures which include faulting, landslides, liquefaction, densification, and ground cracks; (2) tectonic uplift and subsidence; and (3) vibratory ground motion. Seismic hazards originating in marine environment that affect coastal GLFL facilities include tsunamis and seiches (ASCE, 1983).

Siting criteria for nuclear power plants are quite detailed and are governed by regulatory criteria and requirements In general there are no acceptable methods or approaches that are common to the siting of critical facilities that include pipelines, terminals, major transmission facilities, or substations within urban areas (Hall, 1987).

4.4.2.2   Pipelines

Comments on available codes and standards in the technical literature:

All of the gas transmission systems in the area of the 1971 San Fernando earthquake which were constructed in accordance with the requirements of General Order 112-D of CPUC and ASME B 31.8 as design criteria, proved to be earthquake resistant (McNorgan, 1973). Contrary to the reported adequacy of current (1971) structural codes for gas transmission pipelines (McNorgan, 1973), Eguchi, 1984, reported that the behavior of lifelines in the 1971 San Fernando earthquake showed up some glaring hazards and demonstrated the need for new design approaches and requirements for lifeline systems.

The Code of Federal Regulations (49 CFR - 1990), Title 49, Parts 192 and 195 provides minimum safety standards for gas and liquid fuel pipelines in the United States. Nyman, 1987, states that many gas and oil pipeline companies exceed CFR requirements for safety, and they have established operating procedures to handle all types of general emergency situations. The companies consider emergencies resulting from earthquakes not altogether different from those resulting from improper excavation, floods, or fires except that the emergencies may occur simultaneously following an earthquake.

The liquid fuel industry has been using ANSI/ASME B 34.1 for the construction of cross country lines. The gas line companies use a somewhat similar standard, ANSI/ASME B 31.8. Both of these standards have been used by all pipeline industries since the 1950 s. The standards have been reviewed and updated by code committees consisting of industry and non industry members (Bagwell and Marshall, 1973).

Procedures for rational and positive corrosion control have been developed and can be economically applied to underground pipelines. The oil and gas industries, and other major users of underground facilities are developing and using underground corrosion control procedures (Kinsey, 1973).

The current state-of-the-art of computational procedures for pipeline loading and performance under earthquake conditions can not be assimilated into the industry until appropriate design criteria and guidelines are established for the information that these procedures produce (Row, 1987). For the past 10 years, designers who have designed earthquake fault crossings have used strain based design criteria. However, the governing U.S. codes of practice (CFR 192 and 195; ASME/ANSI B 31.4 and B 31.8) do not address such criteria. Foreign codes of practice such as DNV, Rules for Submarine Pipelines, Norway, and the draft Canadian pipeline code (CSA Z 187), recommend the use of inelastic analysis methods and strain-based criteria under special circumstances. Row, 1987, concludes that effort must be directed at including the results of research and current state-of-the-art practice into U.S. codes of practice.

The design factors and stress limits in the ANSI/ASME B 31.4 and B 31.8 codes ensure that certain minimum strength resistances are not exceeded by the effects of specified operational loadings. The code provisions are normally adequate for general stress design of conventional pipelines and piping systems (Price and Barnette, 1987). Conventional systems are considered to be onshore pipelines in stable ground. When it is necessary to evaluate structural discontinuities and dynamic fatigue effects, however, the designer is responsible for determining supplemental local stress design specifications. The alternate rules in ANSI/ASME BPV-VIII (ASME Boiler and Pressure Code) are useful for reference purposes and for guidance in evaluating discontinuity, peak stress fatigue, and plastic cycling fatigue from cyclic loads which are not explicitly accounted for in the ANSI/ASME B 31.4 and B 31.8 standards.

The Public Utilities Commission of the State of California issued the following general order in June 1979, "General Order No. 112-D, Rules Governing Design, Construction, Testing, Maintenance, and Operation of Utility Gas Gathering, Transmission, and Distribution Piping Systems, Liquified Natural Gas Facilities Safety Standards". According to Anderson and Bachman, 1985, the resulting design requirements from the General Order are several times greater than the most demanding building code provisions and are generally more conservative than those specified for nuclear power plants in highly seismic regions. The General Order contains the provisions of: The Uniform Building Code; "Tentative Provisions for the Development of Seismic Regulations for Buildings", ATC 3-06, Applied Technology Council, U.S. Government Printing Office, Wash, D.C.; June 1978; API 650, "Welded Steel Tank for Oil Storage", API 650, 6th.ed., Revision 3, Appendix E, Seismic Design of Storage Tanks, American Petroleum Institute, Wash, DC, Oct. 1979; "Final Safety Analysis Report San Onofre Nuclear Generating Station Units 2 and 3," Southern California Edison Company and San Diego Gas and Electric Company, June 1979.

It is noted that current design practice does not include a method for calculating the stress induced in a pipe due to longitudinal bending, nor does current practice address the potential fatigue of longitudinal and girth welds (O'Rourke et al. 1987).

4.4.2.3   Storage Tanks

Comments on available codes and standards in the technical literature:

Nyman and Kennedy, 1987 note that design standards such as API 650 and AWWA D100 provide a generally adequate approach to the design of liquid storage tanks. Nearly all liquid storage tanks are quite flexible and unanchored. Improved methods that require an estimate of the tank's natural frequency and a design spectral acceleration are available. The choice of an appropriate approach will generally depend upon the allowable stresses outlined in the design criteria and the risks associated with tank damage (Nyman and Kennedy, 1987).

4.4.2.4   Structures and Facilities

Comments on available codes and standards in the technical literature:

Equipment that is vital for maintaining proper pipeline control and initiating emergency actions must remain operational after an earthquake. Nyman and Kennedy, 1987, reported that presently there are no guidelines that apply for the seismic qualification of equipment for oil and gas lifeline systems.

There are also standards for seismic qualification of equipment used in pipeline facilities. One applicable guideline standard is IEEE 1975, Standard 344, Recommended Practices for Seismic Qualification of Class IE Equipment for Nuclear Power Generating Stations (Anderson, 1985). An important part of any seismic qualification programs is to procure equipment of good quality construction and perhaps heavier gauge material than usual, as this could eliminate problems during qualification (Price and Barnette, 1987).

Conventional structures in seismic zones similar to California are designed to current seismic standards according to the Uniform Building Code (UBC) for the appropriate seismic zone. The UBC procedures for the analysis and design are applicable for above ground structures included in pipeline systems (Nymann and Kennedy, 1987).

4.5 Summary

The greatest earthquake-related threat to pipeline fuel transportation systems is from special site hazards such as fault-displacements, liquefaction, landslides, and tsunamis. By judicious siting, this threat can be minimized, but not entirely avoided for pipelines, and can be for the most part avoided for storage facilities, structures and support facilities. Our present state of knowledge in seismology, geology, and geotechnical engineering enables us to plan and design systems in a manner which will minimize exposure to special site hazards, but with the exception of regulations for LNG facilities, siting requirements are not included in present standards and regulations for pipeline transportation systems.

Standards for pipelines include Federal regulations, ASME recommendations, military standards and State of California standards. With the exception of the State of California standards, none of the standards for pipelines address siting requirements. The federal regulations for liquid fuel pipelines, and the ASME standards, which are also adopted by reference in the military standards, contain a requirement for resisting earthquake forces, but do not mention displacements.

Standards for storage tanks include federal regulations for LNG facilities, API and AWWA standards, military standards and State of California regulations. The Federal regulations and the State of California regulations for LNG facilities address siting and secondary storage. These issues are not addressed in any of the other standards. However, the API and military standards have seismic design provisions, addressing lateral forces. Some authors noted that these provisions should be improved.

Standards for structures and support facilities can be derived for the most part from local building codes, model codes, military specifications, and other documents dealing with the design of earthquake-resistant structures. However, there are no standards specifically dealing with the seismic qualification of equipment components, specifically addressed to pipeline transportation systems, and there is probably a need for special provisions for the tiedown of equipment.

5. FEDERALLY CONTROLLED SYSTEMS

5.1 Introduction

Pertinent Federal Agencies were contacted to obtain the following information:

- Oil and gas pipelines operated, leased or regulated by the Agency.

- Requirements for earthquake resistant design and construction for pipelines under the Agency jurisdiction.

- Plans for retrofit of pipelines which are inadequately protected against seismic hazards.

This section contains concise summaries of the information obtained from telephone conversations with persons from the Agencies contacted. Summaries of telephone conversations with the persons contacted from each Agency are given in Appendix A.

37

5.2. Federal Practices

DEPARTMENT OF TRANSPORTATION (DOT)

The DOT regulates pipelines through CFR Title 49 Parts 191 to 193, and 195 for natural and other gas and hazardous liquids respectively. These regulations provide minimum safety standards for the United States and they apply to national pipeline systems owned and operated by pipeline operators. Federally owned pipeline systems are exempt from DOT regulations.

The DOT regulations for pipelines address some natural hazards but do not contain explicit seismic design requirements. State utility commissions generally regulate the distribution piping systems by certifying with DOT to administer that program.

DOT follows general rule making procedures in the preparation and issuing of regulations. Petitions for rule making and notices are sent out so that interested persons and organizations can respond and participate. In addition to its regulatory responsibility, the DOT owns a few pipelines and is responsible for their operation. For example, DOT owns a pipeline system at the St. Lawrence Seaway. With this particular pipeline systems there have been no problems caused by seismic events. However, there have been some leaks. There has been some retrofitting and replacement of pipe due to corrosion. Emergency procedures provide for pipeline shut down.

As an example, DOT recently reviewed a liquid fuel transmission pipeline design for a section of pipeline near San Bernardino, CA that replaced a section of pipeline that was damaged by a derailed train. The regulations used in the design were 49 CFR Part 195 and included seismic provisions including an equation for adding hoop stress and outside stress in accordance with ANSI/ASME B 31.4 and B 31.8. With regard to seismic provisions for the Trans-Alaskan pipeline, sections that cross known faults are above ground. DOT personnel believe that geological studies are performed in many areas to avoid seismic risk, especially on the West Coast. However, in the Northeastern part of the United States seismic design provisions are not usually considered.

Both DOT and the Department of Interior's Minerals Management Service regulate pipeline operators offshore facilities. About 75 percent of these facilities are regulated by DOT and the others are regulated by the Department of Interior.

DEPARTMENT OF ENERGY (DOE)

With the exception of the Federal Energy Regulatory Commission, DOE does not own, operate, lease, or regulate oil, fuel, or gas transmission pipeline systems. They do own, and are responsible for the operation of, lines on their sites.

As an example, DOE is responsible for petroleum reserves stored in California and Wyoming. The Naval Petroleum Reserve became a part of DOE in 1975. In the petroleum reserves, oil is pumped from ships to underground caverns which may involve commercial pipelines. At some petroleum reserve sites, DOE utilizes an operating contractor who also does the engineering work. Each DOE site has to prepare a site development plan.

There have not been any reported earthquake related problems with pipeline systems. However, because of corrosion and modernization of existing lines, there has been some retrofit and replacement of existing pipeline systems. Some post World War II pipeline systems have been replaced with modern equipment. Performance is periodically checked by helicopter overflights over the pipeline system and by monitoring pipelines for corrosion.

The most up-to-date codes and standards are used in the design and construction of pipeline systems. DOE does not have its own set of design guidelines. Contractors must follow DOE orders which include UBC by reference (zone 4). DOE Order 6430.1A (1988) presents design criteria for new facilities and includes safety classes. Also used by DOE is UCRL-15910, "Design And Evaluation Guidelines For DOE Facilities Subject To Natural Phenomenon Hazards". This guideline includes seismic provisions. DOE also has guidelines for nuclear power plants (NE F9-2T). API requirements are followed for pipelines and tanks along with Navy requirements for these systems and also for retrofit and replacement. DOE does not have a retrofit policy. Retrofit is performed on a case by case basis, but is not generally applicable to pipelines. A seismic analysis is always addressed in the design of pipeline systems. These systems, which include all the necessary components, are designed by contractors selected by DOE. Guidelines are needed by DOE for seismic provisions for the design of pipeline systems.

The DOE Office of Energy Emergency Operations has looked into the possible effect of a major earthquake on pipelines in the New Madrid fault area. This effort has been coordinated with those of the Center for Strategic International Studies. The principal concerns are with mitigation, response, and recovery of pipelines subjected to a major earthquake in the New Madrid fault area. Issues to be considered are design criteria, recommendations for operations, needed research, and financial estimates regarding mitigation, response, and recovery. As an example, redundant pipeline systems may be considered as a possible solution to avoid a catastrophe.

FEDERAL ENERGY REGULATORY COMMISSION (FERC)

The FERC was established under DOE by Title 18, CFR, "Conservation of Power and Water Resources", Chapter I. Gas pipelines and LNG facilities are subject to FERC review. Seismic provisions fall under the environmental protection requirements of 18 CFR.

With respect to gas pipelines, FERC reviews proposed designs for potential earthquake hazards where pipelines cross streams, rivers, or geological faults.

For LNG facilities, FERC conducts project-specific reviews of proposed engineering designs using engineering judgments. The designs must comply with 49 CFR Part 193. More specific guidelines for site investigation are spelled out in Section 380.12 (6) and (7) of 18 CFR. NBSIR 84-2833 (Kovacs, 1984) which has been prepared for FERC contains guidelines for site investigations.

FERC seismic and geologic reviews include geologic descriptions of the project area and detailed consideration of adjacent faults, potential for landslides and liquefaction, and area seismicity.

DEPARTMENT OF INTERIOR (DOI)

Under the Outer Continental Shelf (OCS) Lands ACT (43 U.S.C. 1334), the Minerals Management Service (MMS) in the Department of the Interior (DOI) issues leases for the exploration and development of oil and gas and other minerals in the OCS. The MMS issues pipelines rights-of-way on the OCS for the transportation of oil, natural gas, sulphur, or other minerals, under such regulations and conditions as may be prescribed by the Secretary of the Interior (or, where appropriate, the Secretary of Transportation). Regulatory responsibilities of MMS focus on prevention of waste, protection of the environment, conservation of natural resources, production measurement, and safety of OCS lessee and right- of-way holder activities.

There are about 18,000 miles of offshore oil and natural gas transmission pipelines that are jointly regulated by the DOI and the DOT under a 1976 Memorandum of Understanding (MOU) on offshore pipelines. Of the total, 4,500 miles, primarily gathering lines associated with oil and gas production facilities, are regulated solely by MMS. The MMS jurisdiction over offshore pipelines is in the OCS and ends at the Federal\State jurisdictional boundary, generally 3 miles from shore. The DOT's jurisdiction includes pipelines both in the OCS and in State waters.

Under the terms of the 1976 MOU, there are between 70 and 80 structures (primarily pipeline manifold and compressor platforms) currently under DOT jurisdiction. The MOU is being considered for a possible revision in which the jurisdiction of these structures and pipelines would change.

If the effects of scouring, soft bottoms, or other environmental factors are observed to be detrimentally affecting a pipeline, a plan of corrective action must be submitted for MMS approval and following repairs a report of the remedial action taken must be submitted to the MMS by the lessee or right-of-way holder.
During the past 10 years, pipeline-related spills have accounted for about 95 percent by volume of all oil spilled from OCS operations. Therefore, to further reduce spillage related to OCS production, it is necessary to concentrate more on pipeline operations.

During the period 1964 through 1989, of the total volume spilled from pipelines, about 93 percent was from ship-pipeline interactions, primarily ship anchors being dragged over pipelines. This percentage increased to nearly 97 percent for the period 1981-1989. Ship-pipeline interactions usually result in very large spills that heavily skew spill statistics. For example, one such incident in February 1988 resulted in a spill of 14,944 barrels, or about 70 percent of the volume of all spills of 50 barrels or greater for the period 1981 through 1989.

An analysis of 20 years of pipeline failure data compiled by MMS concluded that most remaining pipeline spillage results from pipeline corrosion. Pipeline failures due to external corrosion is more frequent among small sized lines, whereas failures due to internal corrosion are more frequent among medium and larger size pipelines (Mandke, 1990). As the Gulf of Mexico pipeline infrastructure ages, corrosion problems may occur more frequently.

The **Bureau of Land Management** monitors the pipelines across federal lands to determine if the provisions of the right-of-way grant are met. An example of one of the pipelines monitored by the Department of Interior is the Trans-Alaskan pipeline. Design and construction of pipelines across federal lands must comply with DOT regulations.

DEPARTMENT OF DEFENSE (DOD)

The DOD owns some fuel transmission pipeline systems. As an example, they have a pipeline in Maine that is over 200 miles long. There are branch lines from commercial transmission pipelines that provide delivery service to many military installations. DOD also has internal distribution systems at military installations.

In most instances, the distribution systems for natural gas are commercial systems which are locally owned and operated. Essentially the DOD is a customer in the market place and depends on the commercial sector for fuel. They pay for what they use.

Past earthquakes have not caused problems for DOD pipeline systems. Some leaks have occurred in pipelines, but these were due mainly to accidents (or other unforseen conditions or events) or cathodic action on the pipelines. Some earthquake related problems were reported for commercial transmission pipeline systems such as Southern Pacific and Cal Nev. However, specific problems were not identified.

Requirements developed by the Defense Fuel Supplies Center (DFSC) have been used in the design and construction of pipeline systems. Contractors designing pipelines for DOD follow state and federal codes. There have been studies for DOD regarding seismic needs or requirements for retrofit. Many lines have been replaced because of environmental requirements.

DOD - NAVY

The Navy has fuel and oil storage facilities. These facilities are sometimes operated by contract and include transporting oil from docking areas to fuel tanks and from oil storage to barges. The Navy stores fuel for the Air Force and the Army. The Navy does not own, operate, lease, or regulate any oil, fuel, or natural gas transmission pipeline systems. They do have distribution pipeline systems for liquid fuel and for natural gas.

The Navy is responsible for the operation, maintenance, and construction of their pipeline systems. Some storage facilities date back to 1918 and many others were installed during World War II. Many pipeline systems are old. Some pipelines have been replaced because of corrosion and some have been retrofitted. Retrofit has included some internal lining, additional isolation, and automation provided for leak detection. Retrofit by commercial pipeline companies has involved plastic linings in plastic pipes. Only one type of valve was approved by the Navy that was considered to function satisfactorily. This valve for pipe-tank connections has been used for new construction and for replacement.

Past earthquakes caused no significant problems to oil, fuel, or natural gas pipelines. Some water lines were damaged at Treasure Island, California during the Loma Prieta earthquake. There was no damage to utility systems during this earthquake. Liquefaction has been the cause of some pipeline problems. Some buildings were damaged during the Alaskan earthquake, and there was damage to non-Navy tanks during the 1971 San Fernando earthquake. Decisions on shutting down pipelines are made by individual bases.

Standards used in the design and construction of pipeline systems include NAVFAC P-355.1, "Seismic Design Guidelines For Essential Buildings"; NAVFAC Design Manual 22, "Petroleum Fuel Facilities"; API 650 for tanks; and CFR Title 49 Parts 192 and 195. Navy designs are according to seismic zone and Navy guide specs are used. Geological studies are carried out along with a seismic analysis for design of Naval facilities. Borings are generally required. The Navy generally requires more isolation capability than given in standards and guidelines. Double ball joints are used for tank connections. The standards and guidelines available and used by

the Navy do not give much guidance with regard to seismic provisions in the design and construction of oil, fuel, and natural gas pipeline systems. The designer is alerted to address seismic provisions in the design of pipeline systems.

DOD - ARMY

The Army has some pipelines that they own and operate. Many of these pipelines are old. The Army also has some fuel storage tanks. There has not been reported damage to Army-owned pipeline systems caused by earthquakes, however, one person interviewed thought that he had heard of some pipeline system damage attributed to earthquakes.

With regard to design criteria for pipeline systems, the Army uses Technical Manual TM 5-809-10-1, "Seismic Design Guidelines For Essential Buildings"; guide specifications for new construction of gas distribution systems and liquid fuel storage systems; and Navy Manual 22, "Petroleum Fuel Facilities". DOT regulations are generally followed, even though facilities on DOD property are exempt from DOT regulations. None of the above listed criteria address seismic provisions for transmission pipeline systems.

There has been some retrofit of pipeline systems, but not for reasons of seismic damage or to provide earthquake resistant design. Tanks and pipelines have been repaired and in some cases pipelines have been replaced. Each base has its own policy regarding retrofit and replacement procedures.

DOD - AIR FORCE

The Air Force owns and operates liquid fuel pipeline systems located on their installations in the United States and in foreign countries. Some of the pipeline systems are old. The Air Force does not own, operate, lease, or regulate oil, fuel, or natural gas transmission pipeline systems.

Design criteria used by the Air Force for pipeline systems include DOT regulations, API requirements, ASME/ANSI requirements, and Army and Navy guide specs and design manuals. Guide specs for the design of bulk storage tanks are being updated. In general, present design criteria do not address seismic provisions for earthquake resistant design for pipeline systems. Design considerations generally include environmental and safety requirements.

There have not been any known problems with Air Force owned or operated pipeline systems caused by earthquakes. Also, there are not any plans for retrofitting these systems based on seismic considerations. There has been some repair of tanks because of leaks.

The recent (November 1990) design of above - ground steel tanks for aircraft fuel storage for the Air Force was based on criteria for seismic zones 1 and 2. The specification stated that if the site specific design criteria exceed the general design criteria, structural elements shall be redesigned if necessary. Seismic investigation and redesign shall be in accordance with API 650 and TM 5-809-10/NAVFAC P-355/AFM 88-3,Chapter 13. The tanks were designed by a contractor by authority of the Corps of Engineers. Navy and Army guide specifications were used in the design of these fuel storage tanks. Specific provisions for seismic design were not given in the specification for the design of these steel tanks.

GENERAL SERVICE ADMINISTRATION (GSA)

GSA is not involved with oil, fuel, or natural gas transmission pipeline systems. On GSA sites, distribution lines are the responsibility of the utility company. GSA is mostly concerned with buildings and with regard to seismic design provisions they have adopted the UBC. GSA is also currently preparing additional seismic criteria for the design of their buildings.

DEPARTMENT OF HOUSING AND URBAN DEVELOPMENT (HUD)

HUD is not involved with oil, fuel, or natural gas transmission pipeline systems, however they have some involvement with distribution pipeline systems, in particular for public housing. HUD generally provides the funds for these distribution systems and approves the design of the systems. Local codes, reference standards, and reference model building codes are used in the design of the distribution pipeline systems.

NATIONAL AERONAUTICS AND SPACE ADMINISTRATION (NASA)

NASA does not own, operate, lease, or regulate oil, fuel, or natural gas transmission pipeline systems. They do own and operate gas distribution lines at some facilities such as the Ames Research Center, California. Transmission pipelines are operated by utility companies. From these transmission pipelines, NASA takes oil and gas into the facility for internal use. The internal lines range in age from relatively new to the 1940's. The distribution lines include some pipeline support facilities.

There have been minor problems at the Ames Research Center caused by earthquakes such as broken gas lines and leaks at connections and valves. However, the overall performance of the piping systems has been satisfactory. There was an earthquake in 1989 at the Ames Research Center.

Current design procedures which include national standards are used by NASA in the design and construction of pipeline systems. There has been little retrofit or replacement of pipelines. Retrofit for gas lines at the Ames Research Center has been by the use of plastic linings, and steel pipe has been replaced with plastic pipe. Pacific Gas and Electric Company (PGE) uses plastic gas lines in their distribution systems.

Some emergency procedures included use of backup generators with alternate fuel sources for generation of power and heat, and for short periods a propane backup system is available. Pipelines can be shut down or isolated in some cases by acceleration activated automatic valves and some manually operated valves are in use.

OTHER FEDERAL AGENCIES

The **TENNESSEE VALLEY AUTHORITY** (TVA) is not involved with oil, fuel, or natural gas transmission pipeline systems. They do have some distribution systems. The **DEPARTMENT OF AGRICULTURE** (DOA), the **ENVIRONMENTAL PROTECTION AGENCY** (EPA), and the **FEDERAL HIGHWAY ADMINISTRATION** (FHWA) are not involved with oil, fuel, or natural gas transmission pipeline systems.

6. SUMMARY AND RECOMMENDATIONS

6.1 Summary of Findings

6.1.1 System Vulnerability

The overall performance record of gas and liquid fuel pipeline systems subjected to earthquakes has been good. However, serious failures did occur in several earthquakes, particularly in areas of unstable soils.

Modern welded ductile steel pipelines with adequate corrosion protection have a good performance record. Failures, which have occurred in these types of pipelines, were mostly caused by fault displacements, lateral spreading and settlement due to liquefaction, and other large permanent soil displacements. Older pipelines have been damaged many times, including welded pipelines built generally before 1950 in accordance with quality control standards less stringent than those used currently, and segmented cast iron pipelines. Corrosion was the cause of some of the failures that occurred. Pipeline locations vulnerable to damage are bends, elbows, tees, and local eccentricities, and joints in segmented pipelines. Some pipeline failures were attributable to the collapse of supporting structures to which they were attached, such as bridges.

Above-ground storage tanks, particularly those with large height-to-width ratios, have been damaged in many earthquakes. This is attributed to the fact that unpressurized tanks, though structurally very efficient to support vertical gravity loads, are not structurally efficient to resist lateral earthquake forces. The most frequent failure mode is buckling of the shell (elephant's foot buckling). Some of the failures were caused by inadequate anchorage and excessive foundation settlements. Failures were also caused because of inadequate flexibility of pipe connections and contact of the sloshing liquid with the roof structure.

Pumping stations and compressor stations are not very vulnerable. Other above ground support facilities, which were designed to resist earthquakes, suffered only limited damage, which in many instances was caused by inadequate tiedown of equipment or insufficient anchorage to the supporting foundations. Equipment outages were mostly caused by falling debris, collision with other items, sliding, or failure of electrical supplies.

6.1.2 Remedial Measures

The most efficient and economical way to obtain earthquake protection for new facilities is proper siting. Storage tanks and other above-ground facilities should be kept away from special hazard areas, such as geological faults, liquefiable soil deposits, potentially unstable slopes, areas of deep soft soil deposits, and areas that could be subjected to tsunamis, seiches, and other forms of flooding. Pipelines cannot always be located away from hazardous sites, particularly distribution lines, but their exposure to special site hazards can be minimized.

In addition to proper siting, pipelines and structures can be designed to resist earthquake loads and displacements. Criteria for pipeline design were presented by ASCE (1984), and criteria for tanks and other structures are incorporated in many existing standards.

Protection against the environmental consequences of storage tank failures can be provided by secondary containment using earth dikes. Such secondary containments are presently required for LNG storage facilities but not for oil storage tanks.

Connections with tanks, underground utility structures, and valves deserve special consideration. Such connections should be flexible, allowing for rotation to accommodate differential settlement, and should have adequate horizonal restraint to prevent pullout. Penetrations of building walls and bridge abutments should be oversized, with appropriate waterproof and compressible packing to allow for differential settlement. The integrity and ductility of gas, liquid fuel, and critical water pipelines must be ensured at penetrations.

For existing facilities, retrofit and replacement of older facilities in critical areas should be considered. Methods for inspecting and retrofitting older pipelines are available.

Gas and liquid fuel companies operating in areas vulnerable to earthquakes should prepare a formal and formally approved emergency plan for safety, restoration, and environmental protection. The plan should consider a variety of natural disasters applicable within the company's jurisdiction, including seismic events. The plan should address reciprocal agreements for assistance with neighboring utilities, alternative supplies of electricity and water, communications, monitoring, and practices of emergency operations. It is recognized that it is not possible or practical to design and maintain gas and liquid fuel pipeline systems that will never experience an emergency situation. The Federal regulations help establish procedures to deal with emergencies in an orderly fashion (ASCE, 1984).

6.1.3 Existing Guidelines and Standards

Present standards for pipeline systems generally do not address adequately their protection against earthquake damage. This is attributable to the perception that modern welded pipelines are not vulnerable to earthquakes. Current allusions to seismic loading and ground subsidence conditions in CFR Title 49 Parts 192 and 195, as well as relevant ASME standards, are thin and not sufficiently detailed to give reasonable guidance for line pipe and related facilities. Some improvement in the design standards should be sought. Such improvements might ultimately take the form of explicit and carefully phrased performance specifications with reference to relevant detailed design standards which have been developed in model form by organizations such as TCLEE, NIST and NCEER.

Present standards for above-ground liquid fuel storage tanks contain provisions for lateral force design which are based on the present state of the art.

Neither the pipeline standards nor the standards for oil storage tanks address the need for siting studies, even though such studies are often performed in practice. This deficiency could have adverse consequences, particularly in the Central and Eastern U.S., where the need for earthquake resistant design is not always fully recognized. There are also no secondary containment requirements for liquid fuel storage tanks, even if these tanks are located in environmentally sensitive areas.

Standards for LNG storage facilities contain siting criteria, secondary storage provisions and lateral force design requirements.

6.1.4 Federal Practices

Three Federal Agencies have regulatory responsibilities for pipeline fuel transportation systems: The DOT regulates oil and gas pipelines with respect to safety; FERC regulates the fuel transmission rates charged by oil and gas transmission lines, as well as monitors compliance of gas pipelines and all LNG facilities, including terminal and storage facilities, with safety and environmental requirements; and the MMS regulates compliance of offshore production and transmission facilities with safety and environmental requirements. MMS regulates approximately 25 percent of the offshore transmission pipelines, and DOT regulates the other 75 percent. To some extent, the responsibilities of these agencies overlap.

The review and approval of facilities by these agencies are based on the relevant provisions in the Code of Federal Regulations (CFR) and on engineering judgment. Explicit requirements for geological and seismological studies, secondary storage, and earthquake resistant design are included in the federal regulations for LNG facilities. The federal regulations for gas pipelines, as well as other standards referenced in these regulations, do not address earthquake resistant design. The federal regulations for liquid fuel pipelines have a very general requirement for earthquake resistant design which is not adequate for clear guidance on seismic factors. However, ASME and API standards, which address earthquake resistant design for liquid fuel storage tanks, are adopted by reference. These latter standards do not address the siting and secondary storage problems.

Most federal agencies do not own and operate pipeline systems (except for granting rights of way), however some agencies do own and operate relatively short oil and gas transmission pipelines. Many agencies own distribution systems and storage facilities. Most agencies address the earthquake problem in some way, but there is no uniform approach to the problem between agencies, and sometimes within agencies.

6.2 Recommendations

It is recommended that a guideline be prepared for Federal Agencies for earthquake resistant design of gas and liquid fuel pipeline systems which will promote a uniform approach to earthquake resistant practices by all Federal Agencies. This guideline should adopt existing standards and regulations by reference, but add requirements in the following areas:

- Seismologic, geologic, and geotechnical studies and siting requirements for gas and liquid fuel pipeline systems, including storage and support facilities.

- Secondary storage requirements for above-ground oil storage tanks in environmentally sensitive locations.

It is also recommended that the seismic site exploration provisions for LNG facilities given in NBSIR 84-2833 be reviewed and updated if necessary. This document was prepared in 1984, and it should be coordinated with present seismic zoning concepts.

Since the proposed federal guideline may eventually result in an updating of present federal regulations, close coordination between FEMA, DOT, FERC, and MMS, as well as input from industry, will be required.

47

7. ACKNOWLEDGMENT

The authors gratefully acknowledge Dr. Thomas D. O'Rourke for providing important review comments. The library staff at the American Gas Association and the American Petroleum Institute are also acknowledged for their assistance in providing information.

8. REFERENCES

Anderson, James C. and Johnson, Sterling B., "Seismic Behavior of Above-Ground Oil Pipelines", Earthquake Engineering and Structural Dynamics, Vol. 3, No.4, April-June 1975, pp.319-336.

Anderson, James C. and Singh, Anand K., "Seismic Response of Pipelines on Friction Supports," Journal of the Engineering Mechanics Division, Vol.102, No.EM 2, April 1976.

Anderson, T.L., "Seismic Design Considerations for Oil and Gas Pipeline Systems", Proceedings of the 1985 Pressure Vessels and Pipeline Conference, PVP, Vol.98-4, American Society of Mechanical Engineers, 1985, pp.111-117.

Anderson, Thomas L. and Bachman, Robert E., "LNG Terminal Design for California", Journal of the Technical councils of ASCE, Vol.107, No.TC1, American Society of Civil Engineers, April 1985.

Anderson, Thomas L. and Nyman, Douglas J., "Lifeline Engineering Approach to Seismic Qualification", Journal of the Technical Councils of ASCE, Vol.105, No.TC1, American Society of Civil Engineers, April 1979, pp.149-161.

Anderson, Thomas L. and Nyman Douglas J., "Lifeline Earthquake Engineering for Trans Alaska Pipeline System", Proceedings of Conference on the Current State of Knowledge of Lifeline Earthquake Engineering, American Society of Civil Engineers, August 1977, pp.35-49.

ANSI/AWWA D100 - 84 and D100a - 89, Welded Steel Tanks for Water Storage, with 1989 addendum, American Water Works Association, 1989.

ANSI/IEEE Standard 693-1984, "IEEE Recommended Practices for Seismic Design of Substations", Institute of Electrical and Electronic Engineers, 1984.

ANSI/IEEE Standard 344-1987, "IEEE Recommended Practices for Seismic Qualification of Class 1E Equipment for Nuclear Power Generating Stations", Institute of Electrical and Electronic Engineers, 1987.

Anton, Walter F., "Seismic Design of Pumping Plants", Journal of the Technical Councils of ASCE, Vol.107, No.TC1, American Society of Civil Engineers, April 1981, pp.1-12.

API RP 2A, Recommended Practice for Planning, Designing and Constructing Fixed Offshore Platforms, Eighteenth Edition, American Petroleum Institute, September 1989.

API RP 2A-LRFD, Draft Recommended Practice for Planning, Designing and Constructing Fixed Offshore Platforms - Load and Resistance Factor Design, First Edition, American Petroleum Institute, December 1989.

API STD 620, "Design and Construction of Large, Welded, Low-Pressure Storage Tanks", Eighth Edition, American Petroleum Institute, June 1990.

API STD 650, "Welded Steel Tanks for Oil Storage", Eighth Edition, American Petroleum Institute, November, 1988.

API STD 1104, "Welding of Pipelines and Related Facilities", American Petroleum Institute, Washington, DC, Seventeenth Edition, June 1989.

Ariman, T., "A Review of Buckling and Rupture Failures in Pipelines Due to Large Ground Deformations", Proceedings of Symposium on Lifeline Engineering, Earthquake Behavior and Safety of Oil and Gas Storage Facilities, Buried Pipelines and Equipment, PVP-Vol. 77, The American Society of Mechanical Engineers, 1983.

Ariman, Teoman, "Behavior of Buried Pipelines Under Large Ground Deformations in Earthquakes", Proceedings of the US-Japan Workshop on Seismic Behavior of Buried Pipelines and Telecommunications Systems", Tsukuba City, Japan, December 1984.

Ariman, Teoman, "A Review of the Earthquake Response and Aseismic Design of Underground Piping Systems", Proceedings of Conference on the Current State of Knowledge of Lifeline Earthquake Engineering, American Society of Civil Engineers, August 1977, pp.282-292.

Army TM 5-809-10, Navy NAVFAC P-355, Air Force AFM 88-3, Chap. 13, "Seismic Design for Building, February, 1982.

Army TM 5-809-10-1, Navy NAVFAC P-355.1, Air Force AFM 88-3, Chap. 13, Sec. A, "Seismic Design Guidelines for Essential Buildings, February, 1986.

ASCE Committee on Dynamic Analysis of the Committee on Nuclear Structures and Materials of the Structural Division, "The Effects of Earthquakes on Power and Industrial Facilities and Implications for Nuclear Power Plant Design", American Society of Civil Engineers, 1987.

ASCE Technical Council on Lifeline Earthquake Engineering, "Guidelines for the Seismic Design of Oil and Gas Pipeline Systems", Prepared by the Committee on Gas and Liquid Fuel Lines, American Society of Civil Engineers, 1984. (473 pp.).

ASCE Technical Council on Lifeline Earthquake Engineering, "Advisory Notes on Lifeline Earthquake Engineering", American Society of Civil Engineers, 1983.

ASCE, Los Angeles Section, "Earthquake Damage Evaluation and Design Considerations for Underground Structures", Los Angeles, California, February 1974.

ASME (ANSI) B31.4 - 1989, Liquid Transportation System for Hydrocarbons, Liquid Petroleum Gas, Anhydrous Ammonia and Alcohol, American Society of Mechanical Engineers, 1989.

ASME (ANSI) B31.8 - 1989, Gas Transmission and Distribution Piping Systems, American Society of Mechanical Engineers, 1989.

ASME Guide for Gas Transmission and Distribution Piping Systems, American Society of Mechanical Engineers, 1986.

ATC-13,"Earthquake Damage Evaluation Data for California", Applied Technology Council (Funded by Federal Emergency Management Agency), 1985.

Ayala, Gustavo A. and O'Rourke, Michael J., "Effects of the 1985 Michoacan Earthquake On Water Systems and Other Buried Lifelines in Mexico", Technical Report NCEER-89-0009, National Center for Earthquake Engineering Research, State University of New York at Buffalo, March 8, 1989.

Bagwell, Marshall V., "Pipeline Transportation in the 70's", Transportation Engineering Journal of ASCE, Vol.99, No.TE1, American Society of Civil Engineers, February 1973, pp.5-15.

Bettinger, Richard V., "Economic and Seismic Mitigation for Gas and Electric Utility", Proceedings of Conference on Social and Economic Impact of Earthquakes on Utility Lifelines, Edited by J. Isenberg, American Society of Civil Engineers, 1980, pp.98-106.

BOCA, "The BOCA National Building Code", Building Officials and Code Administrators international, Inc., Country Club Hills, IL, 1987.

Building Seismic Safety Council, "Abatement of Seismic Hazards to Lifelines: An Action Plan", FEMA 142, Earthquake Hazards Reduction Series 32, Federal Emergency Management Agency, Washington, DC, August, 1987.

California Public Utility Commission, General Order No.112-D, "Rules Governing Design, Construction, Testing, Maintenance, and Operation of Gas Utility Gathering, Transmission, and Distribution Piping Systems", San Francisco, CA, 1988.

Campbell, Kenneth W., Eguchi, Ronald T. and Duke Martin C., "Reliability in Lifeline Earthquake Engineering", Journal of the Technical Councils of ASCE, Vol.105, No.TC2, December 1979, pp.259-270.

Code of Federal Regulations, 18 CFR, Chapter I, "Federal Energy Regulatory, Commission, Department of Energy", October 1990.

Code of Federal Regulations, 30 CFR, "Mineral Resources", Chapter II, "Mineral Management Service, Department of the Interior", Part 250, "Oil and Gas and Sulphur Operations in the Outer Continental Shelf", October 1990.

Code of Federal Regulations, 49 CFR, "Transportation", Chapter I, "Research and Special Programs Administration"; Part 190, "Pipeline Safety Program Procedures; Part 191, "Transportation of Natural and Other Gas by Pipeline"; Part 192, "Transportation of Natural and Other Gas by Pipeline, Minimum Federal Safety Standards"; Part 193, "Liquefied Natural Gas Facilities; Federal Safety Standards; Part 195, "Transportation of Hazardous Liquids by Pipeline", October 1990.

Combra, F.J., "A Study of Liquid Storage Tank Seismic Uplift Behavior", Proceedings of Symposium on Lifeline Earthquake Engineering, Earthquake Behavior and Safety of Oil and Gas Storage Facilities, Buried Pipelines and Equipment, PVP-Vol. 77, The American Society of Mechanical Engineers, 1983.

Darragh, R.D., "Geotechnical Investigations for Pipelines Subject to Large Ground Deformations", Proceedings of Symposium on Lifeline Earthquake Engineering, Earthquake Behavior and Safety of Oil and Gas Storage Facilities, Buried Pipelines and Equipment, PVP-Vol. 77, The American Society of Mechanical Engineers, 1983.

EERI, "Loma Prieta Earthquake Reconnaissance Report", Supplement to Volume 6, Technical Editor, Lee Benuska, Earthquake Spectra, Journal of the Earthquake Engineering Research Institute, 1990.

EERI, "Reducing Earthquake Hazards: Lessons Learned from Earthquakes", Publication No. 86-02, Earthquake Engineering Research Institute, El Cerrito, CA, November 1986.

Edgar, T.F., Himmelblau, D.M. and Bickel, T.C., "Optimal Design of Gas Transmission Networks", Society of Petroleum Engineers Journal, Vol.18 No.2, AIME, 1978, pp.96-104.

Eguchi, Ronald T., "Seismic Risk to Natural Gas and Oil Systems", FEMA 139, Earthquake Hazard Reduction Series 30, July 1987, pp. 15-33.

Eguchi, Ronald T., "Seismic Risk and Decision Analysis of Lifeline Systems", Proceedings of Symposium on Lifeline Earthquake Engineering : Performance Design and Construction, James D. Cooper, Editor, American Society of Civil Engineers, 1984. pp.82-106.

Eguchi, Ronald T., Hudson, James M., Philipson, Lloyd L., Legg, Mark R., Taylor, Craig and Wiggins, John H., "Earthquake Performance of Natural Gas Distribution Systems", Proceedings of the 1981 International Gas Research Conference, Gas Research Institute, 1982, pp.1761-1770.

Elhmadi, Kamel and O'Rourke, Michael J., "Seismic Damage to Segmented Buried Pipelines", Earthquake Engineering and Structural Dynamics, Vol.19, John Wiley & Sons, Ltd, 1990, pp.529-539.

Factor, Stanley and Grove, Sandra J., "Alaskan Transportation: An Overview of Some Aspects of Transporting Alaskan Crude Oil", Marine Technology, Vol.16, No.3, The Society of Naval Architects and *83* Marine Engineers, July 1979, pp.211-224.

Ford, D.B., "Joint Design for Pipelines Subject to Large Ground Deformations", Proceedings of Symposium on Lifeline Earthquake Engineering, Earthquake Behavior and Safety of Oil and Gas Storage Facilities, Buried Pipelines and Equipment, PVP-Vol, 77, The American Society of Mechanical Engineers, 1983.

Ford, Duane B. "Effects of Ground Failure on Ductile Iron Pipe", Proceedings of Conference on Pipeline Infrastructure, Edited by Bruce A. Bennett, American Society of Civil Engineers, June 6-7,1988, pp.476-487.

Francis, R.F., "Opportunities for Skill-The Management of the National Transmission System", Gas Engineering and Management, Vol.24, No.5, United Kingdom, May 1984.

Fritsche, William F., Jr., "Pipeline for the Future", Pipeline and Gas Journal, Vol.215, No.10, October 1988.

Fu-Lu, M., "Earthquake Response of Fluid-Filled Pipelines Buried in Soil", Proceedings of Symposium on Lifeline Engineering, Earthquake Behavior and Safety of Oil and Gas Storage Facilities, Buried Pipelines and Equipment, PVP-Vol. 77, The American Society of Mechanical Engineers, 1983.

Giuliano, F.A., Editor, "Introduction to Oil and Gas Technology", 2nd Ed., Int. Human Resources Development Corporation, Boston, MA, 1981.

Guan-Qing, C., "Seismic Behavior of the Qinhuangdao /Beijing Oil Pipeline in the Tangshan Earthquake", 80-C2/PVP-84, American Society of Mechanical Engineers, 1980.

Hair, John D. and Hair, Charles W., III, "Considerations in the Design and Installation of Horizontally Drilled Pipeline River Crossings", Proceeding of Conference on Pipeline Infrastructure, Bruce A. Bennett, Editor, Pipeline Division of the Society of Civil Engineers, June 6-7, 1988.

Hale, Dean, "Pipe Renovation", Pipeline and Gas Journal, Vol.211, No.11, September 1984, pp.14-17.

Hall, W.J., "Earthquake Engineering Research Needs Concerning Gas and Liquid Fuel Lifelines", FEMA 139, Earthquake Hazard Reduction Series 30, July 1987, pp. 35-49.

Hall, J.W. and Newmark, N.M., "Seismic Design Criteria for Pipelines and Facilities", Proceedings of Conference on the Current State of Knowledge of Lifeline Earthquake Engineering, American Society of Civil Engineers, August 1977, pp.18-34.

Hall, W.J. and O'Rourke, T.D., "Seismic Behavior and Vulnerability of Pipelines", Proceedings, Third U.S. Conference on Lifeline Earthquake Engineering, Los Angeles, CA, August 1991.

Hall, W.J. and O'Rourke, T.D., "Seismic Behavior and Vulnerability of Lifelines", Proceedings, Third U.S. Conference on Lifeline Earthquake Engineering, TCLEE Monograph No.4, American Society of Civil Engineers, 1991, pp. 761-773.

Hall, William J. and Newmark, Nathan M., "Seismic Design Criteria for Pipelines and Facilities", Journal of the Technical Councils of ASCE, Vol.104, No.TC1 American Society of Civil Engineers, November 1978, pp.91-107.

Hall, William J. and Kennedy, Robert P., "Gas and Liquid Fuel Pipeline Seismic Design", Proceedings of Conference on Social and Economic Impact of Earthquakes on Utility Lifelines", Edited by J. Isenberg, American Society of Civil Engineers, 1980, pp.150-165.

Haroun, M.A., "Assessment of Seismic Hazards to Liquid Storage Tanks at Port Facilities", paper presented at POLA Seismic Workshop, Port of Los Angeles, CA, March 1990, 15p.

Haroun, M.A. and Tayel, M.A., "Numerical Investigation of Axisymmetrical Vibrations of Partly-Filled Cylindrical Tanks", Proceedings of Symposium on Lifeline Earthquake Engineering, Earthquake Behavior and Safety of Oil and Gas Storage Facilities, Buried Pipelines and Equipment, PVP-Vol. 77, The American Society of Mechanical Engineers, 1983.

Haroun, Medhat A., "Behavior of Unanchored Oil Storage Tanks : Imperial Valley Earthquake", Journal of Technical Topics in Civil Engineering, Vol.109, No.1, American Society of Civil Engineers, April 1983, 23-40.

Haroun, Medhat A. and Housner, George W., "Seismic Design of Liquid Storage Tanks", Journal of the Technical Councils of ASCE, Vol.107, No.TC1, American Society of Civil Engineers, April 1981, pp.191-207.

ICBO, "Uniform Building Code", International Conference of Building Officials, Whittier, CA, 1988.

Isenberg, Jeremy, "Role of Corrosion in Water Pipeline Performance in Three U.S. Earthquakes", Proceedings of the Second U.S. National Conference on Earthquake Engineering, Earthquake Engineering Research Institute, August 1979, pp.683-692.

Johnson, W.T., Jr., "Post Earthquake Recovery in Natural Gas Systems-1971 San Fernando Earthquake", Proceedings of Symposium on Lifeline Engineering, Earthquake Behavior and Safety of Oil and Gas Storage Facilities, Buried Pipelines and Equipment, PVP-Vol. 77, The American Society of Mechanical Engineers, 1983.

Katayama, Tsuneo, Kubo, Keizaburo, and Sato Nobuhiko, "Earthquake Damage to Water and Gas Distribution Systems", Proceedings of the First U.S. National conference on Earthquake Engineering, Ann Arbor, MI, 1975.

Kennedy, J.L., Oil and Gas Pipeline Fundamentals, Penn Well Publishing Company, Tulsa OK, 1984, 271p.

Kennedy, R.P., Darrow, A.C. and Short, S.A., "General Considerations for Seismic Design of Oil Pipeline Systems", Proceedings of Conference on the Current State of Knowledge of Lifeline Earthquake Engineering, American Society of Civil Engineers, August 1977, pp.2-17.

Kennedy, Robert P., Darrow, Arthur C. and Short, Stephen A., "Seismic Design of Oil Pipeline Systems", Journal of the Technical Councils of ASCE, Vol.105, No.TC1, American Society of Civil Engineers, April 1979, pp.119-134.

Kennedy, Robert P., Chow, Andrew W. and Williamson, Robert A., "Fault Movement Effects on Buried Oil Pipeline", Transportation Engineering Journal of ASCE, Vol.103, No.TE5, American society of Civil Engineers, September 1977, pp.617-633.

Kinsey, William R.., "Underground Pipe Corrosion", Transportation Engineering Journal of ASCE, Vol.99, No.TE1, American Society of Civil Engineers, February 1973, pp.167-182.68

Kovacs,W.D. et al., NBSIR 84-2833, "Data Requirements for the Seismic Review of LNG Facilities", National Bureau of Standards, June 1984.

Kubo, Keizaburo, Katayama, Tsuneo and Ohashi, Masamitsu, "Lifeline Earthquake Engineering in Japan",,Journal of the Technical Councils of ASCE Vol.105 No.TC1, American Society of Civil Engineers, April 1979, pp.221-238.

Kyriakides, H.D. and Yew, C.H., "Buckling of Buried Pipelines Due to Large Ground Movements", Proceedings of Symposium on Lifeline Earthquake Engineering, Earthquake Behavior and Safety of Oil and Gas Storage Facilities, Buried Pipelines and Equipment, Teoman Ariman, Editor, PVP- Vol.77, The American Society of Mechanical Engineers, 1983.

Lee, B.J. and Ariman, T., "Buckling of Buried Pipeline Crossing An Active Fault", Proceedings of the 1985 Pressure Vessels and Piping Conference, Seismic Performance of Pipelines and Storage Tanks, PVP-Vol.98-4, The American society of Mechanical Engineers, 1985, pp.25-33.

Lee, S.C. and Nash, W.A., "Seismic Behavior of Cylindrical Liquid Storage Tanks Reinforced by Circumferential Rings", Proceedings of Symposium on Lifeline Earthquake Engineering Earthquake Behavior and Safety of Oil and Gas Storage Facilities, Buried Pipelines and Equipment, PVP-Vol. 77, The American Society of Mechanical Engineers, 1983.

Mandke, J.S., "Corrosion Causes Most Pipeline Failures in Gulf of Mexico", Oil and Gas Journal, October 29,1990, pp.40-44.

Manos, G. C. and Clough, R.W., "The Measured and Predicted Shaking Table Response of a Broad Tank Model", Proceedings of Symposium on Lifeline Earthquake Engineering, Earthquake Behavior and Safety of Oil and Gas Storage Facilities, Buried Pipelines and Equipment, PVP-Vol.77, The American Society of Mechanical Engineers, 1983.

Mashaly, E.A. and Datta, T.K., "Seismic Behavior of Buried Pipelines - A State-of-the-Art Review", Journal of Pipelines, Vol.7, No.3, November 1983, pp.215-234.

McCaffrey, M.A. and O'Rourke, T.D., "Buried Pipeline Response to Reverse Faulting During the 1971 San Fernando Earthquake", Proceedings of Symposium on Lifeline Earthquake Engineering, Earthquake Behavior and Safety of Oil and Gas Storage Facilities, Buried Pipelines and Equipment, PVP-Vol. 77, The American Society of Mechanical Engineers, 1983.

McNorgan, John D., "Gas Line Response to Earthquakes", Transportation Engineering Journal of ASCE, Vol.99, No.TE4, American Society of Civil Engineers, November 1973, pp.821-826.

McPartland Brian J., "Innovative Approach Provides Solutions for Crude Oil Pipeline Control", Electrical Construction and Maintenance, Vol.87, No.7, July 1988, pp.72-78.

Meyersohn, W.D. and O'Rourke, T.D., "Pipeline Buckling Caused by Compressive Ground Failure During Earthquakes", Proceedings, Third Japan-U.S. Workshop on Earthquake Resistant Design of Lifeline Facilities and Countermeasures for Soil Liquefaction, Technical Report NCEER-91-0001, NCEER, Buffalo, NY, February 1991, pp.489-496.

Monzon-Despang, H. and Shah, H.C., "Seismic Load Severity Patterns for Lifelines", Proceedings of Symposium on Lifeline Engineering, Earthquake Behavior and Safety of Oil and Gas Storage Facilities, Buried Pipelines and Equipment, PVP-Vol. 77, The American Society of Mechanical Engineers, 1983.

National Petroleum Council, "Petroleum Storage and Transportation", U.S. Department of Energy, April 1989.

NFPA 59 A, "Standard for the Production, Storage, and Handling of Liquefied Natural Gas (LNG)", National Fire Protection Association, Quincy, MA, 1990.

NIST Special Publication 778,"Performance of Structures During the Loma Prieta Earthquake of October 17, 1989," H. S. Lew, Editor, U. S. National Institute of Standards and Technology, January 1990.

NOAA, "San Fernando, California, Earthquake of February 9, 1971", Leonard M. Murphy, Scientific Coordinator, Vol.II, Utilities, Transportation, and Sociological Aspects, U.S. National Oceanic and Atmospheric Administration, 1973.

NAVFAC DM - 22, "Petroleum Fuel Facilities", Design Manual 22, Naval Facilities Engineering Command, August 1982.
Nyman, Douglas J., "Operations and Maintenance Considerations for Mitigation of Earthquake Effects on Oil and Gas Pipelines", FEMA 139, Earthquake Hazard Reduction Series 30, July 1987, pp. 1-13.

Nyman, Douglas J. and Kennedy, Robert P., "Seismic Design of Oil and Gas Pipeline Systems", FEMA 139, Earthquake Hazard Reduction Series 30, July 1987, pp. 51-66.

Ogawa, N., "A Vibration Test of Earthquake Induced Hydraulic Transients of Liquid-Filled Pipelines", Proceedings of Symposium on Lifeline Engineering, Earthquake Behavior and Safety of Oil and Gas Storage Facilities, Buried Pipelines and Equipment, PVP-Vol.77, The American Society of Mechanical Engineers, 1983.

O'Rourke, Michael J. and Ayala, Gustava, "Seismic Damage to Pipeline : Case Study", Journal of Transportation Engineering, Vol.116, No.2, American Society of Civil Engineers, March/April 1990, pp.123-134.

O'Rourke, Michael J., "Response Analysis of Crude Oil Transmission Systems", Structural Safety and Reliability, Proceedings of ICOSSAR '89/5th Int'l Conf., San Francisco, CA, August 7-11, 1989.

O'Rourke, Michael J., "Approximate Analysis Procedures for Permanent Ground Deformation Effects on Buried Pipelines", Proceedings 2nd U.S. Japan Workshop on Liquefaction, Large Ground Deformation and Their Effects on Lifelines, Buffalo, N.Y., September 1989 (NCEER Report No. NCEER-89-0032).

O'Rourke Michael J., "Mitigation of Seismic Effects on Water Systems", Seismic Design and Construction of Complex Civil Engineering Systems, Symposium Sponsored by TCLEE/ASCE, National Convention, St. Louis, MO, October 27, 1988, pp.65-78.

O'Rourke, T.D., "Geotechnical Considerations for Buried Pipelines", FEMA 139, Earthquake Hazard Reduction Series 30, July 1987, pp. 67-82.

O'Rourke, T.D., Ingraffea, A.R., Norman, R.S. and Burnham, K.B., "Evaluation of Cased and Uncased Gas Pipelines at Railroads", Proceedings of the 1986 International Gas Research Conference, Published by Government Inst. Inc., Rockville, MD, 1987, pp.286-297.

O'Rourke, T.D., Grigoriu, M.D. and Khater, M.M., "Seismic Response of Buried Pipelines", Pressure Vessel and Piping Technology 1985 A Decade of Progress, American Society of Mechanical Engineers, 1985, pp.281-323.

O'Rourke, T.D., Roth, B., Miura, F. and Hamada, M., "Case History of High Pressure Pipeline Response to Liquefaction-Induced Ground Movements", Proceedings, 4th U.S. National Conference on Earthquake Engineering, Palm Springs, CA, May 1990, Vol.1, pp.955-964.

O'Rourke, T.D. and Tawfik, M.S., "Effects of Lateral Spreading on Buried Pipelines During the 1971 San Fernando Earthquake", Proceedings of Symposium on Lifeline Earthquake Engineering, Earthquake Behavior and Safety of Oil and Gas Storage Facilities, Buried Pipelines and Equipment, PVP-Vol. 77, The American Society of Mechanical Engineers, 1983.

O'Rourke, T.D. and Trautmann C.H., "Earthquake Ground Rupture Effects on Jointed Pipe", Proceedings Second Specialty Conference of the Technical Council on Lifeline Earthquake Engineering, D.J. Smith, Editor, 1981, pp.65-80.

Price, P. StJ. and Barnette, J.A., "A Pipeline Code and Structural Criteria For Pipelines in Arctic and Earthquake Regions", Conference Proceedings: Offshore and Arctic Pipelines-1987, American Society of Mechanical Engineers, 1987, pp.89-97.

Row, Dennis, "Computational Methods for Analysis of the Response of Buried Pipelines to Soil Movements and Ground Distortion", FEMA 139, Earthquake Hazard Reduction Series 30, July 1987, pp. 83-104.

Saito, K., Nishio, N. and Katayoma, T., "Recommended Practice for Earthquake Resistant Design of Medium and Low Pressure Gas Pipelines", Proceedings of Symposium on Lifeline Engineering, Earthquake Behavior and Safety of oil and Gas Storage Facilities, Buried Pipelines and Equipment, PVP-Vol. 77, The American Society of Mechanical Engineers, 1983.

SBCCI, "Standard Building Code", Southern Building Code Congress international, Inc., Birmingham, AL, 1988.

Schiff, A.J., Yanev, M.P.I. and Givin, H.H., "Retrofitting Lifeline Facilities : The Japanese Approach", Proceeding of Symposium on Lifeline Earthquake Engineering : Performance, Design and Construction, American Society of Civil Engineers, 1984, pp.107-116.

Schiff, Anshell and Yanev, Peter, "Performance of Lifeline Systems", Earthquake Spectra, August 1989, pp.114-135.

SEAOC, Seismology Committee, "Recommended Lateral Force Requirements and Commentary", Sacramento, CA, 1990.

Shaoping, S., Zongpei, A., and Ganyl, H., "Pipeline Damage and Its Relationship with Joints", Proceedings of Symposium on Lifeline Engineering, Earthquake Behavior and Safety of Oil and Gas Storage Facilities, Buried Pipelines and Equipment, PVP-Vol. 77, The American Society of Mechanical Engineers, 1983.

Shibata, H., Shigeta, T., Kubota, M., and Morikawa, S., "On Some Results on Response Observation of Liquid Storage Tanks to Natural Earthquake," Proceedings of Symposium on Lifeline Earthquake Engineering, Earthquake Behavior and Safety of Oil and Gas Storage Facilities, Buried Pipelines and Equipment, PVP-Vol. 77, The American Society of Mechanical Engineers, 1983.

Singhal, A.C., "Axial, Bending and Torsional Behavior of Pipelines", Journal of Technical Topics in Civil Engineering, Vol.110, No.1, American Society of Civil Engineers, May 1984, pp.38-47.

Singhal, A.C. and Benavides, J.C. "Pull Out and Bending Experiments in Buried Pipes", Proceedings of Symposium on Lifeline Engineering, Earthquake Behavior and Safety of Oil and Gas Storage Facilities, Buried Pipelines and Equipment, PVP-Vol. 77, American Society of Mechanical Engineers, 1983.

Singhal, Avi, "How to Design Pipelines for Earthquake Resistance-1", Pipeline and Gas Journal, Vol.210, No.8, July 1983, pp.32,34,36-37.

Singhal, Avi, "How to Design Pipelines for Earthquake Resistance III," Pipeline and Gas Journal, Vol.210, No.11, September 1983.

Takada, S., "Experimental Study on Mechanical Behavior of PVC Pipelines Subjected to Ground Subsidence", Proceedings of Symposium on Lifeline Engineering, Earthquake Behavior and Safety of Oil and Gas Storage Facilities, Buried Pipelines and Equipment, PVP-Vol. 77, The American Society of Mechanical Engineers, 1983.

Tobin, Thomas L., "California's Urban Hazards Mapping Program: A Bold Experiment in Earth Science and Public Policy", Seismic Safety Commission, 1900 K Street, # 100, Sacramento, CA, 1991.

Toki, K., Fukumori, Y., Sako, M. and Tsubakimoto, T., "Recommended Practice for Earthquake Resistant Design of High Pressure Gas Pipelines", Proceedings of Symposium on Lifeline Engineering, Earthquake Behavior and Safety of Oil and Gas Storage Facilities, Buried Pipelines and Equipment, PVP-Vol. 77, The American Society of Mechanical Engineers, 1983.

Trautmann, C.H. and O'Rourke, T.D., "Load-Displacement Characteristics of A Buried Pipe Affected by Permanent Earthquake Ground Movements", Proceedings of Symposium on Lifeline Engineering, Earthquake Behavior and Safety of Oil and Gas Storage Facilities, Buried Pipelines and Equipment, PVP-Vol. 77, The American Society of Mechanical Engineers, 1983.

TRB Committee for Public Safety, "Pipelines and Public Safety", Special Report 219, Transportation Research Board, National Research Council, Washington, DC, 1988.

Tsai, Robert J., Gordon, Edward, Simpson, Kenneth O. and Olson, Robert R., "Optimal Gas pipeline Design Via Dynamic Programming With Variable Stages," PSIG Annual Meeting (New Orleans, Louisiana), Pipeline Simulation Interest Group, October 1986.

Wang, L.R.L., "Role and Development of Soil Parameters for Seismic Responses of Buried Lifelines", Proceedings of Symposium on Lifeline Engineering, Earthquake Behavior and Safety of Oil and Gas Storage Facilities, Buried Pipelines and Equipment, PVP-Vol. 77, The American Society of Mechanical Engineers, 1983.

Wang, Leon Ru-Liang and O'Rourke, Michael J., "Overview of Buried Pipelines Under Seismic Loading, Journal of the Technical Councils of ASCE, Vol.104, No.TC1 American Society of Civil Engineers, November 1978, pp.121-130.

Wang, Leon R.L., "Seismic Counter Measures for Buried Pipelines", Proceedings of Conference on Pipeline Infrastructure, Edited by Bruce A. Bennett, American Society of Civil Engineers, June 6-7,1988, pp.502-517.

Ward, Delbert B. and Taylor, Craig F., "Formulating Utah's Seismic Policy for Lifelines", Proceedings of Conference on Social and Economic Impact of Earthquakes on Utility Lifelines, Edited by J. Isenberg, American Society of Civil Engineers, 1980, pp.202-214.

Weidlinger, Paul and Nelson, Ivan, "Seismic Design of Underground Lifelines", Journal of the Technical Councils of ASCE, \vol.106, No.TC1, American Society of Civil Engineers, August 1980, pp.185-200.

Wenzel, A.B., Esparza, E.D. and Westine, P.S., "Pipeline Response to Buried Explosive Detonations", Proceedings of Symposium on Lifeline Engineering, Earthquake Behavior and Safety of Oil and Gas Storage Facilities, Buried Pipelines and Equipment, PVP-Vol. 77, The American Society of Mechanical Engineers, 1983.

Wosniak, R.S. and Mitchell, W.W., Basis for Seismic Design Provisions for Welded Steel Storage Tanks, Proc. refining Department, API, Vol. 57, pp 485-501, American Petroleum Institute, 1987

Yeh, Y-H. and Wang, L.R-L., "Combined Effects of Soil Liquefaction and Ground Displacement to Buried Pipeline", Proceedings of the 1985 Pressure Vessels and Piping Conference, PVP, Vol.98-4, American society of Mechanical Engineers, 1985, pp.43-51.

Yokel, F.Y., FEMA-202, NISTIR 89 - 4213, "Earthquake Resistant Construction of Electric Transmission and Telecommunication Facilities Serving the Federal Government", FEMA Earthquake Hazards Reduction Series 56, September, 1990.

Young, F. M. and Pardon, J.M., "Hydraulic Transients in Liquid-Filled Piping Networks Due to Seismic Excitation", Proceedings of Symposium on Lifeline Engineering, Earthquake Behavior and Safety of Oil and Gas Storage Facilities, Buried Pipelines and Equipment, PVP-Vol. 77, The American Society of Mechanical Engineers, 1983.

DEPARTMENT OF TRANSPORTATION (DOT)

Transportation: **Donald R. Trilling**
Office Of Pipeline Safety: **Cesar Deleon, Bob Holter**
Office of Emergency Transportation: **Pete Sill**

The DOT regulates pipelines through CFR Title 49 Parts 192 and 195 for natural and other gas and hazardous liquids respectively. These regulations provide minimum safety standards for the United States and they apply to national pipeline systems owned and operated by pipeline operators. Federally owned pipeline systems are exempt from DOT regulations.

The DOT regulations address some natural hazards but do not provide specific provisions for earthquake resistant design or seismic design requirements. State utility commissions generally regulate distribution piping systems by certifying with DOT to administer these programs.

DOT follows general rule making procedures in preparing and issuing of regulations. Petitions for rule making and notices are sent out so that interested persons and organizations can respond and participate.

The DOT owns a few pipelines and is responsible for operating them. For example, DOT owns a pipeline system at the St. Lawrence Seaway. There have been no problems with the pipeline systems caused by earthquakes. There have been some leaks and some retrofitting and replacement of pipe due to corrosion. Emergency procedures provide for pipeline shut down.

Bob Holter reviewed a liquid fuel transmission pipeline design for a section of pipeline that was replaced. This section of pipeline, near San Bernardino, CA, was damaged by a derailed train. The regulations used in the design were 49 CFR Part 195 and included seismic provisions including an equation for adding hoop stress and outside stress in accordance with ASME B 31.4 and B 31.8. With regard to seismic provisions for the Trans-Alaskan pipeline, sections that cross known faults are above ground.

Mr.Holter believes that in general geological studies are performed to avoid seismic risk especially on the West Coast. The Northeastern part of the United States is not usually considered for seismic design provisions.

With regard to offshore pipeline facilities, DOT, in conjunction with the Department of Interior's Mineral Management Service, regulates these pipeline operators. About 75 percent of these facilities are regulated by DOT and the others are regulated by the Department of Interior.

DEPARTMENT OF ENERGY (DOE)

DOE : **James R. Hill, Chester Bigelow**
Oak Ridge National Laboratory (ORNL): **Jim Beavers**
Energy Emergency Operations: **Jack Wagner**
Naval Petroleum Reserves: **Mike Ruiz**

DOE does not own, operate, lease, or regulate oil, fuel, or gas transmission pipeline systems. They do own, and are responsible for, the operation of lines on their sites. As an example, DOE is responsible for petroleum reserves stored in California and Wyoming. The Naval Petroleum Reserve became a part of DOE in 1975. With regard to petroleum reserves, oil is pumped from ships to caverns which may involve commercial pipelines. At some petroleum reserve sites, DOE utilizes an operating contractor who also does the engineering work. Each DOE site has to prepare a site development plan.

There have not been any reported problems to pipeline systems caused by earthquakes. However, due to corrosion and modernization, there has been some retrofit and replacement of pipeline systems on site. Some post World War II pipeline systems have been replaced with modern equipment. Routine helicopter flight over the pipeline system and monitoring of pipelines for corrosion are methods for periodic checks on performance.

The most up-to-date codes and standards are used in the design and construction of pipeline systems. DOE does not have its own set of design guidelines. Contractors must follow DOE orders which include UBC (zone 4). DOE Order 6430.1A (1988) contains general design criteria for new facilities and includes safety classes. Also used by DOE is UCRL-15910, "Design And Evaluation Guidelines For DOE Facilities Subject To Natural Phenomenon Hazards". This latter guideline includes seismic provisions. DOE also has guidelines for nuclear power plants (NE F9-2T). API requirements are followed for pipelines and tanks along with Navy requirements for these systems and also for retrofit and replacement. DOE does not have a retrofit policy. Retrofit is performed on a case by case basis, but is not generally applicable to pipelines. A seismic analysis is always used in the design of pipeline systems. These systems which include the necessary components are designed by contractors selected by DOE. Guidelines are needed by DOE for seismic provisions for the design of pipeline systems.

The DOE Office of Energy Emergency Operations has looked into the possible effect of a major earthquake on pipelines in the New Madrid fault area. This effort has been coordinated with those of the Center for Strategic International Studies. The principal concerns are with mitigation, response, and recovery of pipelines subjected to a major earthquake in the New Madrid fault area. Issues to be considered are design criteria, recommendations for operations, needed research, and financial estimates regarding mitigation, response, and recovery. As an example, redundant pipeline systems may be considered as a possible solution to avoid a catastrophe.

FEDERAL ENERGY REGULATORY COMMISSION (FERC)

Bob Arvedlund, Marty Burless, Rich Hoffman,

For oil transmission pipelines, FERC only regulates rates.

For gas pipeline systems, FERC performs the following functions:

- Regulation of rates for gas pipelines
- Authorization to build (environmental issues considered)

Justification of need and intended use of the pipeline is submitted to FERC, and approval must be granted. Requirements for approval, including environmental requirements are given in 18 CFR. The type of data that must be supplied to FERC is also stated in the CFR.

FERC conducts environmental reviews for gas and LNG facilities. For pipelines, the review covers earthquakes to some extent, with particular attention to stream and river crossings (the design is reviewed). With regard to LNG facilities, the design must comply with the DOT requirements in 49 CFR Part 193. More specific guidelines for site investigation are spelled out in Section 380.12 (6) and (7) of 18 CFR. NBSIR 84-2833 (Kovacs, 1984) which has been prepared for FERC contains guidelines for site investigations, and the applicant must supply a report in accordance with the stipulations in NBSIR 84-2833.

FERC reviews are project specific. The review of proposed engineering designs is based on engineering judgment.

DEPARTMENT OF INTERIOR

Bureau of Land Management: **Bob Lawton**
Bureau of Reclamation: **Bob McDonald, David P. Prosser**
Minerals Management Service: **Elmer Danenberger**
USGS: **E.V.Leyendecker, James Devine**
Under the Outer Continental Shelf (OCS) Lands Act (43 U.S.C. 1334), the Minerals Management Service (MMS) in the Department of the Interior (DOI) issues leases for the exploration and development of oil and gas and other minerals in the OCS. The MMS currently regulates about 3,800 oil and gas production platforms and structures on the OCS. This regulation also applies to the pipelines associated with these facilities. The MMS issues pipelines rights-of-way on the OCS for the transportation of oil, natural gas, sulphur, or other minerals, under such regulations and conditions as may be prescribed by the Secretary of the Interior (or, where appropriate, the Secretary of Transportation). Regulatory responsibilities of MMS focus on prevention of waste, protection of the environment, conservation of natural resources, production measurement, and safety of OCS lessee and right-of-way holder activities.

There are about 18,000 miles of offshore oil and natural gas transmission pipelines that are jointly regulated by the DOI and the DOT under a 1976 Memorandum of Understanding (MOU) on offshore pipelines. Of the total, 4,500 miles (primarily gathering lines associated with oil and gas production facilities) are regulated solely by MMS. The MMS jurisdiction over offshore pipelines is in the OCS and ends at the Federal\State jurisdictional boundary, generally 3 miles from shore. The DOT's jurisdiction includes pipelines both in the OCS and in State waters.

Under the terms of the 1976 MOU, there are between 70 and 80 structures (primarily pipeline manifold and compressor platforms) currently under DOT jurisdiction. The MOU is being considered for a possible revision in which the jurisdiction of these structures and pipelines would change.

If the effects of scouring, soft bottoms, or other environmental factors are observed to be detrimentally affecting a pipeline, a plan of corrective action must be submitted for MMS approval and following repairs a report of the remedial action taken must be submitted to the MMS by the lessee or right-of-way holder.

During the past 10 years, pipeline-related spills have accounted for about 95 percent by volume of oil spilled from OCS operations. Therefore, to further reduce spillage related to OCS production, it is necessary to concentrate more on pipeline operations.

During the period 1964 through 1989, of the total volume spilled from pipelines, about 93 percent was from ship-pipeline interactions, primarily ship anchors being dragged over pipelines. This percentage increased to near 97 percent for the period 1981-1989. Ship-pipeline interactions usually result in very large spills that heavily skew spill statistics. For example, one such incident in February 1988 resulted in a spill of 14,944 barrels, or about 70 percent of the volume of all spills of 50 barrels or greater for the period 1981 through 1989.

An analysis of 20 years of pipeline failure data compiled by MMS concluded that most remaining pipeline spillage results from pipeline corrosion. Pipeline failures due to external corrosion is more frequent among small sized lines, whereas failures due to internal corrosion are more frequent among medium and larger size pipelines (Mandke, 1990). As the Gulf of Mexico pipeline infrastructure ages, corrosion problems may occur more frequently.

The Bureau of Land Management monitors the pipelines across federal lands to determine if the provisions of the right-of-way grant are met. An example of one of the pipelines monitored by the Department of Interior is the Trans-Alaskan pipeline. The design and construction of pipelines crossing federal lands must comply with the DOT regulations.

The USGS is currently working on spectral mapping of peak ground accelerations and response spectra for use in seismic design. The USGS is not currently conducting work pertaining to lifelines.

DEPARTMENT OF DEFENSE (DOD)

Defense Fuel Supply Center (DFSC): **Chet Doberson, Anne Scheulan, Eddie French**
Military Transportation Management Command (MTMC): **David Fuchs**
Defense Logistic Agency (DLA): **Don Neri**

The DOD owns some fuel transmission pipeline systems. As an example, they have a pipeline in Maine that is over 200 miles long. There are branch lines off of commercial transmission pipelines that provide delivery service to many military installations. DOD also has internal distribution systems at military installations. In general, the distribution systems for natural gas are commercial systems which are locally owned and operated. Essentially the DOD is a customer in the market place and depends on the commercial sector for fuel. They pay for what they use.

Past earthquakes have not been a problem with the DOD pipeline systems. Some leaks have occurred in pipelines, but these were due mainly to accidents (or other unforseen conditions or events) or cathodic action on the pipelines. There were reported problems with some commercial transmission pipeline systems such as Southern Pacific and Cal Nev which were caused by earthquakes. Specific problems were not identified.

Requirements developed by DFSC have been used in the design and construction of pipeline systems. Contractors designing pipelines for DOD follow state and federal codes. There have been studies for DOD regarding seismic needs or requirements for retrofit. Many lines have been replaced because of environmental requirements.

DOD - NAVY

Naval Facilities Command (NAVFAC): **Howard Nickerson, Harry Zimmerman, Richard Thomas**

Naval Civil Engineering Laboratory (NCEL): **Ting Lee Lew, John M. Ferritto, Gary Anguiano**

The Navy has fuel and oil storage facilities. These facilities are sometimes operated by contract and include transporting oil from docking areas to fuel tanks and from oil storage to barges. The Navy stores fuel for the Air Force and the Army. The Navy does not own, operate, lease, or regulate any oil, fuel, or natural gas transmission pipeline systems. They do have distribution pipeline systems for liquid fuel and for natural gas.

The Navy is responsible for the operation, maintenance, and construction of their pipeline systems. Some storage facilities date back to 1918 and many others were installed during World War II. Many pipeline systems are old. Some pipelines have been replaced because of corrosion and some have been retrofitted. Retrofit has included some internal lining, additional isolation, and automation provided for leak detection. Retrofit by commercial pipeline companies has involved plastic linings in plastic pipes.

There has been little or no problems to oil, fuel, or natural gas pipelines caused by earthquakes. Some water lines were damaged at Treasure Island, California during the Loma Prieta earthquake. There was no damage to utility systems during this earthquake. Liquefaction has been the cause of some pipeline problems. Some buildings were damaged during the Alaskan earthquake and there was damage to non-Navy tanks during the 1971 San Fernando earthquake. Policy decisions on the shutting down of pipelines are made by individual bases.

Standards used in the design and construction of pipeline systems include NAVFAC P-355.1,"Seismic Design Guidelines For Essential Buildings"; NAVFAC Design Manual 22,"Petroleum Fuel Facilities"; API 650 for tanks; and CFR Title 49 Parts 192 and 195. Navy designs are according to seismic zone and Navy guide specs are used. Geological studies are

65

carried out along with a seismic analysis for design of Naval facilities. Borings are generally required. The Navy generally requires more isolation capability than given in standards and guidelines, double ball joints are used for tank connections.

The standards and guidelines available and used by the Navy do not give much guidance with regard to seismic provisions in the design and construction of oil, fuel, and natural gas pipeline systems. The designer is alerted to address seismic provisions in the design of pipeline systems. Only one type of valve was approved by the Navy that was considered to function satisfactorily. This valve for pipe-tank connections has been used for new construction and for replacement.

DOD - ARMY

Office, Chief of Engineers (OCE): **Eward C.Pritchett, Page Johnson, Dale Otterness**

The Army has some pipelines that they own and operate. Many of these pipelines are old. The Army also has some fuel storage tanks. There has not been reported damage to pipeline systems caused by earthquakes, however, one person interviewed thought that he had heard of some pipeline system damage attributed to earthquakes.

With regard to design criteria for pipeline systems, the Army uses Technical Manual TM 5-809-10-1,"Seismic Design Guidelines For Essential Buildings"; guide specifications for new construction of gas distribution systems and liquid fuel storage systems; and Navy Manual 22,"Petroleum Fuel Facilities". DOT regulations are generally followed although DOD is exempt for facilities on DOD property. None of these criteria address seismic provisions for transmission pipeline systems.

There has been some retrofit of pipeline systems, but not for reasons of seismic damage or to provide earthquake resistant design. Tanks and pipelines have been repaired and in some cases pipelines have been replaced. Each base has their own policy regarding retrofit and replacement procedures.

DOD - AIR FORCE

Ronald Wong, Sid McCard

The Air Force owns and operates liquid fuel pipeline systems located on their installations in the United States and in foreign countries. Some of the pipeline systems are old. The Air Force does not own, operate, lease, or regulate oil, fuel, or natural gas transmission pipeline systems.

Design criteria used by the Air Force for pipeline systems include DOT regulations, API requirements, ASME/ANSI requirements, and Army and Navy guide specs and design manuals. Guide specs for the design of bulk storage tanks are being updated. In general, design criteria do not address seismic provisions for earthquake resistant design for pipeline systems. Design considerations generally include environmental and safety requirements.

There have not been any known problems with Air Force owned or operated pipeline systems caused by earthquakes. Also, there are no plans for retrofitting these systems based on seismic considerations. There has been some repair of tanks because of leaks.

Recently designed (November 1990) aircraft fuel storage above-ground steel tanks for the Air Force were based on criteria for seismic zones 1 and 2. The specification stated that if the site specific design criteria exceed the general design criteria, structural elements shall be redesigned if necessary. Seismic investigation and redesign shall be in accordance with API 650 and TM 5-809-10/NAVFAC P-355/AFM 88-3,Chapter 13. The tanks were designed by a contractor by authority of the Corps of Engineers. Many Navy and Army guide specifications were used in the design of these fuel storage tanks. Specific provisions for earthquake resistant design were not given in the specification for the design of the steel tanks.

GENERAL SERVICES ADMINISTRATION (GSA)

Tom Graves, Bruce Hall

GSA is not involved with oil, fuel, or natural gas transmission pipeline systems, they are not in the business of moving fuel. With regard to GSA sites, distribution lines are the responsibility of the utility company. GSA is mostly concerned with buildings and with regard to seismic design provisions they have adopted the Uniform Building Code. They are also currently preparing additional seismic criteria for the design of their buildings.

DEPARTMENT OF HOUSING AND URBAN DEVELOPMENT (HUD)

Davis White, Bob Fuller, Jerry Tobias

HUD is not involved with oil, fuel, or natural gas transmission pipeline systems, however they have some involvement with distribution pipeline systems, in particular with public housing. HUD generally provides the funds for these distribution systems and approves the design of the systems. Local codes, reference standards, and reference model building codes are used in the design of the distribution pipeline systems.

NATIONAL AERONAUTICS AND SPACE ADMINISTRATION (NASA)

NASA: **Charlie Pittinger**
Ames Research Center, CA: **Bob Dolci, Mike Falarski**

NASA does not own, operate, lease, or regulate oil, fuel, or natural gas transmission pipeline systems. They do own and operate gas distribution lines at some facilities such as the Ames Research Center in California. Transmission pipelines are operated by utility companies. From these transmission pipelines, NASA takes oil and gas into the facility for internal use. The internal lines range in age from relatively new to the 1940's. The distribution or internal lines include some pipeline system components.

There have been minor problems at the Ames Research Center caused by earthquakes such as broken gas lines and leaks at connections and valves. In general, the piping systems performed satisfactorily. There was an earthquake in 1989 at the Ames Research Center.

67

Current design procedures which include national standards are used by NASA in the design and construction of pipeline systems. There has been little retrofit or replacement of pipelines. Retrofit for gas lines at the Ames Research Center has been by the use of plastic linings, and steel pipe has been replaced with plastic pipe. PGE uses plastic gas lines in their distribution systems.

Some emergency procedures have included backup generators for alternate fuel sources for generation of power and heat and for awhile there was available a backup propane system. Pipelines can be shut down or isolated in some cases by acceleration activated automatic valves and there are in use some manually operated valves.

TENNESSEE VALLEY AUTHORITY (TVA)

Jerry Cook, Don R. Denton

The TVA is not involved with oil, fuel, or natural gas transmission pipeline systems. They do have some distribution systems.

DEPARTMENT OF AGRICULTURE (DOA)

DOA: Keith Surdiek, James R. Talbot
Rural Electrification Administration: Lee A. Belfore

The Department does not own, operate, lease, or regulate oil, liquid fuel, or natural gas transmission pipeline systems. The only pipelines utilized are those for water and sewage systems, including those for irrigation. They are involved in the design of these systems.

ENVIRONMENTAL PROTECTION AGENCY (EPA)

Thomas J. Moran

The EPA does not own, operate, lease, or regulate oil, liquid fuel, or natural gas transmission pipeline systems.

FEDERAL HIGHWAY ADMINISTRATION (FHWA)

James Cooper

The FHWA is not involved with oil, liquid fuel, or natural gas transmission pipeline systems.

Made in the USA
San Bernardino, CA
08 July 2013